Doing Action Research in Your Own Organization

Doing Action Research in Your Own Organization

Third edition

David Coghlan and Teresa Brannick

$SAGE

Los Angeles | London | New Delhi
Singapore | Washington DC

SAGE Publications Ltd
1 Oliver's Yard
55 City Road
London EC1Y 1SP

SAGE Publications Inc.
2455 Teller Road
Thousand Oaks, California 91320

SAGE Publications India Pvt Ltd
B 1/I 1 Mohan Cooperative Industrial Area
Mathura Road, New Delhi 110 044

SAGE Publications Asia-Pacific Pte Ltd
3 Church Street
#10-04 Samsung Hub
Singapore 049483

Library of Congress Control Number available

British Library Cataloguing in Publication data

A catalogue record for this book is available
from the British Library

ISBN 978-1-84860-215-1
ISBN 978-1-84860-216-8 (pbk)

Typeset by C&M Digitals (P) Ltd, Chennai, India
Printed and bound in Great Britain by the MPG Books Group
Printed on paper from sustainable resources

Contents

About the Authors

David Coghlan is an associate professor at the School of Business, University of Dublin, Trinity College, Dublin, Ireland where he teaches organization development and action research and participates actively in both communities internationally. He has an MSc in management science from Manchester University (UK), an SM in management from the Massachusetts Institute of Technology's Sloan School of Management, a PhD from the National University of Ireland and an MA from the University of Dublin. He is co-author of *Organization Change and Strategy* (Routledge, 2006), and co-editor of *Managers Learning in Action* (Routledge, 2004) and a four-volume set, *The Fundamentals of Organization Development* (Sage, 2010). He is a member of the editorial advisory board of several journals, including *Action Research, The Journal of Applied Behavioral Science, Systemic Practice and Action Research* and *The Journal of Management Education*.

Teresa Brannick is a lecturer in the business research programme at the Michael Smurfit Graduate School of Business at University College, Dublin, Ireland. Her undergraduate degree is in mathematics, her master's in sociology and her PhD in marketing research. She has been a practising researcher for over 20 years and has published over 30 research papers in such diverse fields as epidemiology, public policy, industrial relations and marketing. She is the editor-in-chief of *Irish Journal of Management*. She is co-editor of *Business Research Methods: Theories, Techniques and Sources* (Oak Tree Press, Dublin, 1997). She conducts seminars on research methods and on researching your own organization.

Preface

The theory and the practice of insider action research (IAR) have been afforded increasing attention since the publication of the first edition of this book in 2001. Although Bill Torbert produced in 1976 what appears to have been the first book working from an explicit insider action research approach, that milestone was not marked and that innovative book did not open the door to the emergence of a literature on doing action research in one's own organization. Indeed, any form of insider research has been viewed with suspicion (Brannick and Coghlan, 2007). In our view, the first edition of this book appeared to give voice to a practice that was struggling for legitimacy and which hitherto had not been framed in a manner that facilitated its place in the action research literature. In the intervening years since 2001, further articulations of theory and accounts of practice have burgeoned through the second edition, the special issue of *Action Research* on insider action research (Coghlan and Holian, 2007) and an increasing number of published articles and book chapters, not to mention the number of dissertations in schools of education, healthcare, business and social work that have not found their way into published articles.

In his notion of 'innovation action research', Kaplan (1998) presents an action research cycle of (1) observing and documenting practice, (2) teaching and speaking about it, (3) writing articles and books, (4) implementing the concept and (5) moving to advanced implementation. After several years of observing and documenting practice, teaching and speaking about it, writing articles and books, and thereby implementing the concept, we see this third edition as advanced implementation. The theory and practice of insider action research has advanced, and this third edition seeks to mark a further moment in innovation action research.

What is action research? As the name suggests, action research is an approach to research which aims at both taking action and creating knowledge or theory about that action. The outcomes are both an action and a research outcome, unlike traditional research approaches which aim at creating knowledge only. Action research works through a cyclical process of consciously and deliberately (1) planning, (2) taking action and (3) evaluating the action, leading to further planning and so on. The second dimension of action research is that it is collaborative, in that the members of the system which is being studied participate actively in the cyclical process. This contrasts with traditional research where members are objects of the study. Action research is a generic term that covers many forms of action-oriented research, which may be confusing to any prospective researcher. At the same time, the array of approaches indicates diversity in theory and practice among action researchers and provides a wide choice for potential action researchers as to what might be appropriate for their research.

Action research is appropriate when the research topic is an unfolding series of actions over time in a given group, community or organization, and the members wish to study their own action in order to change or improve the working of some aspects of the system, and study the process in order to learn from it. Hence, action research is related to experiential learning and reflective practice.

We are all insiders of many systems – our families, communities and organizations – and the knowledge we have of these systems is rich and complex. We act to shape the progress of these systems, for example, through parenting and through enacting professional and managerial roles. Doing research in and on one's own organization means that a member of an organization undertakes an explicit research role in addition to the normal functional role that they hold in the organization. The researcher then has to balance their organizational role, which they usually hope will continue, with the additional demands of a role of inquiry and research. Insider action researchers need to be aware of how their roles influence how they view their world as well as how they are perceived by others, and to be able to make choices as to when to step into and out of each of the multiple roles they hold.

There are many issues to be considered for those embarking on research in their own organization or part thereof. From the perspective of individuals who are seeking to do the research in order to achieve academic certification, there are issues of gaining access and receiving permission, and building and maintaining support from peers and relevant subsystems within the organization. There are issues of selecting a research question and area for study. In such a case, student-researchers, in effect, take on a role additional to their conventional organizational one, that of active agent of inquiry and change. This multiple role identity both complicates and focuses the research project. There are issues around how to attain some sense of objectivity and move beyond a personal perspective by testing assumptions and interpretations. There are the uses of appropriate frameworks for viewing and understanding the data. There are questions about how to write up such a research project, give feedback to one's superiors and peers, and disseminate the research to the wider community. Handling interpretations or outcomes which would be perceived negatively by the organization is a particularly sensitive issue. They are also likely to have access to 'external' academic supervisors who advise and support them throughout their research project. However, not all insider action researchers are students; they may be internal consultants or managers, in which case there may not be a planned end point for carrying dual roles, or access to advice and support from key external 'critical friends' to help them sustain doing the work and maintaining a career.

Who does action research in their own organization? A common context for such research is where an individual employee undertakes research as part of an academic programme in order to fulfil requirements for academic certification. In

this instance the individual initiates the research agenda and attempts to negotiate a research project which will meet both her own and the organization's needs.
This occurs in full-time and part-time programmes, at doctorate, master's, undergraduate and diploma levels, and in business, healthcare, government, education,
social work, and third sector organizations. Some research projects may be integrally linked to inquiry into the processes of problem resolution; others may take
a broader, more comprehensive and more diagnostic perspective. At the same
time, selection of a research topic from one's own organization is typically
attached to an expectation or a contract that the research will make a useful contribution to the organization.

Action researchers have to deal with emergent processes, not as distractions but
as central to the research process. The desire to be involved in or to lead radical
change involves high hassle and high vulnerability, which requires a combination
of self-reflection with vulnerability, realistic expectations, tolerance, humility,
self-giving, self-containment and an ability to learn.

Insider action research is an exciting, demanding and invigorating prospect that
contributes considerably to researchers' own learning and contributes to the
development of the systems in which we work and live and with which we have
affiliations. It is also daunting, with a high potential for self-destruction, particularly if roles and politics are not managed well. So what does it take to do insider
research? Our learning from our own work and the work of those managers we
supervise has enabled us to gain some insight into the attributes, competencies
and methodological tools that support effective practice.

Readership

This book is addressed to the reader who is in this dual role of simultaneously
holding an organizational functional role which is linked to a career path and
ongoing membership of the organization, and a more temporary researcher role
for the duration of the research project. While this may imply a distinction
between research and ordinary life, we do not intend such a distinction. Our aim
is to provide a book which is useful for those who select an action research role
in their own organization for a temporary period, and for those in academic insti-
tutions who supervise such research.

There are many books that explore the theory and practice of action research
(Greenwood and Levin, 2007; Stringer, 2007; Reason and Bradbury, 2008; Shani
et al., 2008). We do not intend retracing what is well presented in these works,
particularly with regard to epistemological issues, the history of action research
and detailed formats of research interventions. Indeed, we recommend that this
book be used in conjunction with such works as Greenwood and Levin (2007),
Stringer (2007), Gummesson (2000), Reason and Bradbury (2008) and Shani
et al. (2008).

Plan of the book

The book is divided into three parts and follows a series of questions. We anticipate that these questions are indicative of the ones you have at different stages of your project and that they guide your progress. Part I, 'Foundations', introduces and explores foundational material on action research. Chapter 1 begins from action and poses the question, 'How do you position your action research initiative?' Chapter 2 is built around the question, 'How do you learn in action?' Chapter 3 is grounded in the questions, 'How do you understand what you are doing in your insider action research initiative?' and 'How do you understand how it fits into a rich research tradition?'

Part II, 'Implementation', deals with issues of putting your action research project into action. The question underpinning Chapter 4 is, 'How might you frame and select an action research project?' Chapter 5 answers the questions, 'How do you design your insider action research project?' and 'How do you implement it?' Chapter 6 explores the questions, 'how do you understand the complex interaction of individuals and teams in organizations and in your action research project?' 'How do you work with individuals, with a team and across teams in your project?' Chapter 7 opens with the question, 'How do I use frameworks to make sense of what I am seeking to understand?' and explores the use of frameworks in sense making.

Part III, 'Issues and Challenges in Researching Your Own Organization', deals with issues particular to doing action research in your own organization. Chapter 8 poses the questions, 'What are the implications of engaging in action research in your own organization?' and 'What are the particular dynamics that accompany insider action research that you need to understand and to take into account?' It outlines four different forms which insider research can take, depending on the system's and your own explicit commitment to learning in action. Chapter 9 explores the questions, 'How do you build on the closeness that you have to the organization and maintain distance?' and 'How do you balance the potential dilemmas and tugs between your established organizational roles and your researcher role?' Chapter 10 poses the critical questions, 'How do I survive and thrive in a political environment?' and 'How do I act politically in a mode within the ethics of action research?' Finally, Chapter 11 provides some hints on writing an action research dissertation.

David Coghlan
Teresa Brannick
Dublin

Acknowledgements

We are grateful to those action research colleagues and friends who have provided valuable support, feedback, ideas and materials for the revision: Vivienne Brady, Mary Casey, Anne Coughlan, Paul Coughlan, Rosalie Holian, Geralyn Hynes, Tyler Korb, Ann Martin, Ed Schein, Rami Shani and Bill Torbert. We thank specifically colleagues who shared their work with us and gave us permission to reproduce material from their work: David Besancon, Bob Dick, Arthur Freedman, Eddie Grande, Joel Harmon, Lyle Yorks and Ortrun Zuber-Skerritt.

We are very grateful for the feedback from the many groups who used both previous editions in courses and seminars and from whom we have received feedback. We acknowledge the feedback from successive cohorts of the Masters in Management Practice programme at the Irish Management Institute, University of Dublin, the Masters in Health Service Management at the University of Dublin and the MBA at the Smurfit Graduate School of Business, University College, Dublin, as they contributed to refining some of the frameworks while they struggled through their own insider action research projects. We also received helpful comments through working with doctoral groups at Benedictine University in the USA.

We acknowledge the invaluable help and support of the Sage editorial and production teams, especially Patrick Brindle, Anna Coatman and Ian Antcliff.

Part I

Foundations

ONE

Introducing Action Research

We begin by focusing on the big picture of your insider action research initiative. As you are poised to embark on this venture, we pose the following questions for your consideration. Out of what experience of your organization are you inquiring? What is or has been going on that puzzles you and that you wish to address? What patterns of deliberate action may support this inquiry and provide the opportunity for insights? With whom will you work so that this initiative will be good for you, be worthwhile for your organization and be useful for others? In this chapter we present a brief introduction to action research and then go straight into how it might be enacted. The theoretical discussion of action research and its tenets may be found in Chapter 3.

As a preliminary activity to focus your mind, take a sheet of paper and explore the following questions. Think of a project within your organization on which you are working. What is the context of this project? What is its purpose? What are your hoped-for outcomes? With whom are you working on this project to complete it successfully? What would you need to do if you wanted to understand this project more deeply and thoroughly than if you were simply focusing on solving a problem or completing it as a project? If you were invited to give a talk on your project to interested others, who have not been directly involved, what might you want to say?

This preliminary activity provides an introductory sense of what action research involves. Now begin reading the chapter and those following it to broaden and deepen your understanding and application of action research in your own organization.

A brief introduction to action research

In the words of Reason and Bradbury, 'action research is a participatory, democratic process concerned with developing practical knowing in the pursuit of worthwhile human purposes, grounded in a participatory worldview'

(2008: 1). This working definition provides a flavour of the broad scope and intent of action research with the ultimate aim of 'the flourishing of individual persons and their communities'. Shani and Pasmore provide a more restricted definition:

> Action research may be defined as an emergent inquiry process in which applied behavioural science knowledge is integrated with existing organizational knowledge and applied to solve real organizational problems. It is simultaneously concerned with bringing about change in organizations, in developing self-help competencies in organizational members and adding to scientific knowledge. Finally, it is an evolving process that is undertaken in a spirit of collaboration and co-inquiry. (1985: 439)

Given the context of this book, where we expect readers to be working on action research projects in their own organizations, we are working more from Shani and Pasmore's definition than from Reason and Bradbury's.

Shani and Pasmore present a complete theory of the action research process in terms of four factors (Figure 1.1):

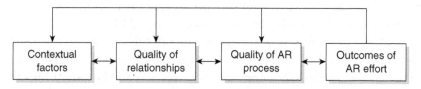

Figure 1.1 Complete theory of action research (Shani and Pasmore, 1985: 444)

- *Context.* These factors set the context of the action research project. Individual goals may differ and impact the direction of the project, while shared goals enhance collaboration. Organizational characteristics, such as resources, history, formal and informal organizations and the degrees of congruence between them, affect the readiness and capability for participating in action research. Environmental factors in the global and local economies provide the larger context in which action research takes place.
- *Quality of relationships.* The quality of relationships between members and researchers is paramount. Hence the relationships need to be managed through trust, concern for others, equality of influence, common language and so on.
- *Quality of the action research process itself.* The quality of the action research process is grounded in the dual focus on both the inquiry process and the implementation process.
- *Outcomes.* The dual outcomes of action research are some level of sustainability (human, social, economic, ecological) and the development of self-help and competencies out of the action and the creation of new knowledge from the inquiry.

Several broad characteristics define action research:

- research *in* action, rather than research *about* action
- a collaborative democratic partnership
- research concurrent with action
- a sequence of events and an approach to problem solving.

We will discuss each in turn.

First, action research focuses on research *in* action, rather than research *about* action. The central idea is that action research uses a scientific approach to study the resolution of important social or organizational issues together with those who experience these issues directly. Action research works through a cyclical four step process of consciously and deliberately (1) planning, (2) taking action and (3) evaluating the action, (4) leading to further planning and so on.

Second, AR is a collaborative, democratic partnership. Members of the system that is being studied participate actively in the cyclical process outlined above. Such participation contrasts with traditional research where members of the system are subjects or objects of the study. An important qualitative element of action research is how people participate in the choice of research focus and how they engage in the processes of action and inquiry.

Third, action research is research concurrent with action. The goal is to make that action more effective while simultaneously building up a body of scientific knowledge.

Finally, action research is both a sequence of events and an approach to problem solving. As a sequence of events, it comprises iterative cycles of gathering data, feeding them back to those concerned, jointly analysing the data, jointly planning action, taking joint action and evaluating jointly, leading to further joint data gathering and so on. As an approach to problem solving, it is an application of the scientific method of fact-finding and experimentation to practical problems requiring action solutions and involving the collaboration and cooperation of the action researchers and members of the organizational system. The desired outcomes of the action research approach are not just solutions to the immediate problems but important learning from outcomes both intended and unintended, and a contribution to scientific knowledge and theory.

Three audiences/voices/practices

An integrative approach to research incorporates three voices and audiences: the first, second and third persons (Reason and Torbert, 2001; Reason and Bradbury, 2008). Traditionally, research has focused on the third person – researchers doing research on third persons and writing a report for other third persons. In a more complete vision of research as presented by action research and many other transformational inquiry approaches, authentic third person research integrates first and second person voices. *First person* research is typically characterized as a form of inquiry and practice that one does on one's own and so addresses the ability of the individual to foster an inquiring approach to his or her own life, to act out of awareness and purposefully. First person research can take researchers 'upstream' where they inquire into their basic assumptions, desires, intentions and philosophy of life. It can also take them 'downstream' where they inquire into their behaviour, ways of relating and action in the world. *Second person* inquiry/practice addresses their ability to inquire into and work with others on issues of mutual concern, through face-to-face dialogue, conversation and joint action. Second person practice poses an important challenge as to who is involved in the research and how. As action research is integrally collaborative and democratic, the quality of second person inquiry and action is central. *Third person* inquiry/practice aims at creating communities of inquiry, involving people beyond the direct second person action. The third person is impersonal and is actualized through dissemination by reporting, publishing and extrapolating from the concrete to the general. As Reason and Torbert (2001) point out, there are plenty of implicit examples of first, second and third person inquiry, but what is required now is explicit integrating of all three persons with action and inquiry. The construct of first, second and third person inquiry is a development of Reason and Marshall's popular notion of three audiences of research:

> All good research is for *me*, for *us*, and *for them*: it speaks to three audiences ... It is *for them* to the extent that it produces some kind of generalizable ideas and outcomes ... It is *for us* to the extent that it responds to concerns for our praxis, is relevant and timely ... [for] those who are struggling with problems in their field of action. It is *for me* to the extent that the process and outcomes respond directly to the individual researcher's being-in-the-world. (1987: 112–13)

Action research approaches are radical to the extent that they advocate replacement of existing forms of social organization. Action research challenges normal science in several action-oriented ways. Sharing the power of knowledge production with the researched subverts the normal practice of knowledge and policy development as being the primary domain of researchers and policy-makers. Action researchers work on the epistemological assumption that the purpose of academic research and discourse is not just to describe, understand and explain the world but also to change it (Reason and Torbert, 2001). The issue is not so much the form of the knowledge produced or the methodology employed to gather data/evidence, but who decides the research agenda in the first place and who benefits directly from it.

The contrast of roles is between that of detached observer in positivist science and that of an actor and agent of change in action research (Evered and Louis, 1981). Weisbord (1988) explores the images of taking photographs and making films in relation to organization development. He describes taking photographs as freezing a moment in time and arranging key factors in a conceptual framework. No photograph takes in the whole of reality; it only takes in what is intended to be included in the frame. Photographers decide what is to be in the frame and they manipulate the setting to include and exclude desirable and undesirable features. In contrast, making films is an engagement in patterns of activity and relationships by multiple actors who are moving and interacting over a period of time and across locations. It is increasingly common to find actors directing their own films. In these cases, actor-directors engage in their acting role in costume and then return to behind the camera in order to study the take, to critique it and to make decisions about proceeding to the next take. We find this image of making films and the action researcher as an actor-director pertinent and useful for thinking about doing action research. As Riordan expresses it, action research projects represent

> a kind of approach to studying social reality without separating (while distinguishing) fact from value; they require a practitioner of science who is not only an engaged participant, but also incorporates the perspective of the critical and analytical observer, not as a validating instance but as integral to the practice. (1995: 10)

In a seminal article, Chandler and Torbert (2003) explore how first, second and third person voice and practice may be engaged in the past, present and future. Much of what we refer to as qualitative research is focused on the past. Action research builds on the past and takes place in the present with a view to shaping the future. Action research scholars, such as Chris Argyris, Edgar Schein, Bill Torbert, Judi Marshall and Peter Reason, have led the way in exploring how to do first person and second person inquiry in the present and how this practice may be rigorous and have quality. Their work permeates this book.

Enacting action research cycles

In its original Lewinian and simplest form, the action research cycle comprises a pre-step and three core activities: planning, action and fact-finding (Lewin, 1946/1997). The pre-step involves naming the general objective. Planning comprises having an overall plan and a decision regarding what the first step to take is. Action involves taking that first step, and fact-finding involves evaluating the first step, seeing what was learned and creating the basis for correcting the next step. So there is a continuing 'spiral of steps, each of which is composed of a circle of planning, action and fact-finding about the result of the action' (Lewin, 1946/1997: 146).

These core steps have been articulated differently by different authors, from Stringer's (2007) simple *look, think, act* to French and Bell's (1999) complex action research organization development (OD) framework involving iterative cycles of joint action planning, feedback, further data gathering, diagnosis and action of an external OD consultant with a client system.

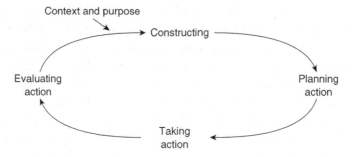

Figure 1.2   The action research cycle

The action research cycles

For the context of doing action research in your own organization we are presenting an action research cycle comprising a pre-step (context and purpose) and four basic steps: constructing, planning action, taking action, and evaluating action (Figure 1.2). The exploration of the action research cycle needs to be understood in terms of the four factors of action research that we presented earlier: context, quality of relationships, quality of the action research process itself, and the outcomes.

Pre-step: context and purpose

The action research cycle unfolds in real time and begins with seeking an understanding of the context of the project. Why is this project necessary or desirable? In terms of assessing the external context, what are the economic, political and social forces driving change? In terms of internal forces, what are the cultural and structural forces driving change? The assessment of these forces identifies their source, their potency and the nature of the demands they make on the system. Included also is the assessment of the degree of choice in how the system responds to the forces for change. Once a sense of the need or desirability for the project is identified then the most useful focus for attention is the definition of a desired future state. The process of defining the desired future state is critical as it sets the boundaries for the purpose of the project and helps provide focus and energy for the later stages. The issues are elaborated in Chapter 5.

Another critical consideration in this pre-step is establishment of collaborative relationships with those who have ownership or need to have ownership of the above questions. A central second person task in this regard is to develop the group or groups with which you will be working on the project.

Main steps

Constructing

In the previous editions of this book, we used the term 'diagnosis' for this step and we are replacing it now with 'constructing'. Contemporary organization development writers have noted that there is a move away from the 'modernist' notion of objectivist diagnosis, with its connotations from medical practice, where data collection is structured prior to any action being taken (Schein, 1999; Bushe and Marshak, 2008). The assumption underpinning this approach is that there is a complex system to be diagnosed, into which interventions may be made with a desired outcome of improvement or transformation. What is replacing this assumption is the assumption that organizations are socially co-constructed and comprise multiple meanings so that there is no single truth to be discovered and no one right way to organize that is independent of the people who make up any particular organization (Campbell, 2000). Accordingly, we are reframing the first step of the action research cycle as a dialogic activity in which the stakeholders of the project engage in *constructing* what the issues are, however provisionally, as a working theme, on the basis of which action will be planned and taken. As this dialogic step involves the articulation of the practical and theoretical foundations of action, it needs to be done carefully and thoroughly. While this constructing may change in later iterations of the action research cycle, any changes in constructing need to be recorded and articulated clearly, showing how events have led to alternative meaning and showing the evidence and rationale for the new shared meanings on which further action is based. It is important that the constructing step be a collaborative venture, that is, that you as the action researcher engage relevant others in the process of constructing and not be the expert who decides apart from others. In Chapter 4 we focus on how a project may be framed, and in Chapter 7 we outline some guidelines for using frameworks for understanding organizational phenomena.

Planning action

Planning action follows from the exploration of the context and purpose of the project, the constructing of the issue, and is consistent with that. It may be that this action planning focuses on a first step or a series of first steps. In Chapter 5 we will describe how you implement the action research project. Again we emphasize the importance of collaboration in planning action.

Taking action

Then the plans are implemented and interventions are made collaboratively.

Evaluating action

The outcomes of the action, both intended and unintended, are examined with a view to seeing:

- if the original constructing fitted
- if the actions taken matched the constructing
- if the action was taken in an appropriate manner
- what feeds into the next cycle of constructing, planning and action.

So the cycle continues (Figure 1.3).

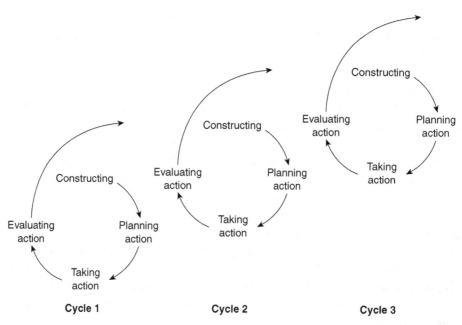

Figure 1.3   Spiral of action research cycles

In any action research project there are multiple action research cycles operating concurrently. These cycles typically have different time spans. The image of a clock captures this usefully (Figure 1.4). The hour hand, which takes 12 hours to complete its cycle, may represent the project as a whole. In a large complex project, it may take several years to complete its cycle. The minute hand, which takes an hour to complete its cycle, may represent phases or particular sections of the project. The second hand, which completes its cycle in a minute, may represent specific actions within the project, e.g. a specific meeting or interview. As

Figure 1.4   Concurrent cycles of action research

with the clock, where the revolutions of the three hands are concurrent, and where the revolutions of the second hand enable the revolutions of the minute hand, and the revolutions of the second and minute hands together enable the completion of the hour hand, the short-term action research cycles contribute to the medium-term cycles which contribute to the longer-term cycle.

While the action research cycle expresses the core process of integrating action and theory, it is important to keep it in perspective. For instance, Heron (1996) describes two approaches to the use of the cycle. He contrasts one approach, *Apollonian*, whereby the cycles are enacted in a rational, linear, systematic manner, with *Dionysian*, an approach where there is an imaginative, expressive, tacit approach to integrating reflection and action. He cautions against being rigid in adapting the action research cycle formally and so denying spontaneity and creativity. It is also important not to get too preoccupied in the cycles at the expense of the quality of participation.

Meta learning

In any action research project there are two action research cycles operating in parallel. This is particularly true where the action research is undertaken for academic accreditation. One cycle is the cycle we have just described of constructing, planning, taking action and evaluating in relation to the achievement of the project's aims. Zuber-Skerritt and Perry (2002) call this the *core* action research cycle. The second cycle is a reflection cycle which is an action research cycle about the action research cycle. Zuber-Skerritt and Perry call this the *thesis* action research cycle. In other words, at the same time as you are engaging in the project or core action research cycles, you need to be constructing, planning, taking action and evaluating around how

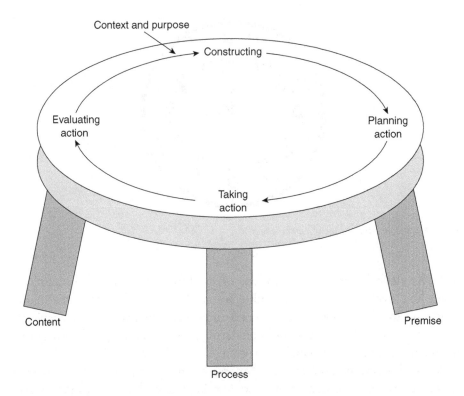

Figure 1.5 Meta cycle of action research

the action research project itself is going and what you are learning. You need to be continually inquiring into each of the four main steps, asking how these steps are being conducted and how they are consistent with each other and, so, shaping how the subsequent steps are conducted. As Argyris (2003) argues in making the same point, this inquiry into the steps of the cycles themselves is central to the development of actionable knowledge. It is the dynamic of this reflection on reflection that incorporates the learning process of the action research cycle and enables action research to be more than everyday problem solving. Hence it is learning about learning – in other words, meta learning.

Mezirow (1991) identifies three forms of reflection: content, process and premise. These are useful categories. *Content* reflection is where you think about the issues, what you think is happening, etc. *Process* reflection is where you think about strategies, procedures and how things are being done. *Premise* reflection is where you critique underlying assumptions and perspectives. All three forms of reflection are critical.

When content, process and premise reflections are applied to the action research cycle, they form the meta cycle of inquiry (Figure 1.5). The *content* of what is constructed, planned, acted on and evaluated is studied and evaluated.

The *process* of how constructing is undertaken, how action planning flows from that constructing and is conducted, how actions follow and are an implementation of the stated plans, and how evaluation is conducted are critical foci for inquiry. There is also *premise* reflection, which is inquiry into the unstated, and often non-conscious, underlying assumptions which govern attitudes and behaviour, such as might be embedded in language. For instance, the culture of the organization or subculture of the group working on the project has a powerful impact on how issues are viewed and discussed, without members being aware of them (Schein, 2004).

If you are writing a dissertation, then the meta cycle is the focus of your dissertation. Remember, the action research project and your dissertation project are not identical. They are integrally interlinked, but they are not the same. The project on which you are working may go ahead irrespective of whether or not you are writing a dissertation. Your dissertation is an inquiry into the project; hence you need to describe both cycles in a way that demonstrates the quality of rigour of your inquiry.

Mezirow's forms of reflection parallel the four territories of experience commonly used in action research (Fisher et al., 2000; Torbert and Associates, 2004). These four territories operate at the individual, interpersonal and organizational levels.

- *Intentions*: purpose, goals, aims and vision
- *Planning*: plans, strategy, tactics, schemes
- *Action*: implementation, performance
- *Outcomes*: results, outcomes, consequences and effects.

Action research aims to develop awareness, understanding and skills across all these territories. You try to understand your intentions, to develop appropriate plans and strategies, to be skilled at carrying them out, to reflect on how well you have carried out the plans, and to evaluate their results. You can also inquire about the connections between these phases. You might, for example, begin with the outcomes, and explore how your actions caused these outcomes. Or you may take the inquiry further, and look at how your intentions and plans shaped your actions.

The activities of the meta cycle are not confined to your first person practice as the individual action researcher. To add another layer of complexity to the learning cycle, the second person practice with the groups and teams engaged in the action research cycles also attends to the steps of content, process and premise reflection.

Attending to the action research cycle and to the meta cycle may involve more than simply attending to behaviour. You may draw from techniques in the qualitative research approaches through how you formulate the issue, collect and analyse data and report results (Sagor, 2004). Techniques from grounded theory approaches may be useful once the core compatibilities and incompatibilities between the two approaches are recognized (Baskerville and Pries-Heje, 1999).

Quality and rigour in action research

The action research paradigm requires its own quality criteria. Action research should be judged *not* by the criteria of positivist science, but rather within the criteria of its own terms. Reason (2006) points to what he considers to be choice points and questions for quality in action research:

1 Is the action research explicit in developing a praxis of relational participation? In other words, how well does the action research reflect the cooperation between the action researcher and the members of the organization?
2 Is action research guided by a reflexive concern for practical outcomes? Is the action project governed by constant and iterative reflection as part of the process of organizational change or improvement?
3 Does action research include a plurality of knowing which ensures conceptual-theoretical integrity, extends our ways of knowing and has a methodological appropriateness? Action research is inclusive of practical, propositional, presentational and experiential knowing and so as a methodology is appropriate to furthering knowledge on different levels.
4 Does action research engage in significant work? The significance of the project is an important quality in action research.
5 Does the action research result in new and enduring infrastructures? In other words, does sustainable change come out of the project?

Reason argues that as an action researcher you need to be aware of these choices and make them clear and transparent to yourself and to those with whom you are engaging in inquiry and to those to whom you present your research in writing or presentations. The editorial guidelines for the journal *Action Research* invite potential contributors to address these dimensions explicitly in submitting their work to the journal.

Rigour in action research refers to how data are generated, gathered, explored and evaluated, how events are questioned and interpreted through multiple action research cycles. In other words, as the action researcher, you need to show:

1 How you engaged in the steps of multiple and repetitious action research cycles (how constructing, planning, taking action and evaluating were done) and how these were recorded to reflect that they are a true representation of what was studied
2 How you challenged and tested your own assumptions and interpretations of what was happening continuously through the project by means of content, process and premise reflection, so that your familiarity with and closeness to the issues are exposed to critique

3 How you accessed different views of what was happening which probably produced both confirming and contradictory interpretations

4 How your interpretations are grounded in scholarly theory, rigorously applied, and how project outcomes are challenged, supported or disconfirmed in terms of the theories underpinning those interpretations and judgements.

What does a good action research project look like? Eden and Huxham (1996) provide an extensive list of the 15 characteristics of good action research. The foundational characteristics reflect the intentionality of the researcher to change an organization, that the project has some implications beyond those involved directly in it, and that the project has an explicit aim to elaborate or develop theory as well as be useful to the organization. Theory must inform the design and development of the actions. Eden and Huxham place great emphasis on the enactment of the action research cycles, in which systematic method and orderliness are required in reflecting on the outcomes of each cycle and the design of the subsequent cycles.

In our view, a good action research project contains three main elements: a good story, rigorous reflection on that story, and an extrapolation of usable knowledge or theory from the reflection on the story. These can be put in terms of three questions. What happened? How you do make sense of what happened? So what?

What happened?

As action research is about real-time change, its core is the story of what takes place. The action research cycle of the general objective pre-step, and the four main steps of constructing, planning, action and fact-finding, describe how the project is conceived, what is intended, the cycles of action and the outcomes, both intended and unintended. The story must be presented in a factual and neutral manner, that is to say, as if it had been recorded on camera, and so that all the actors could agree on what had taken place. In short, the story is based on directly observable behaviour. Therefore, you need to be able to present evidence to support your narrative. Recorded data in journals and organizational documentation are important supporting evidence.

Accordingly, it is critical that fact be clearly distinguished from value, that the basic story does not contain the author's inferences or interpretations, or at least, not without such inferences or interpretations being explicitly identified as such. For instance, if an action research story contains an assertion that a certain group was out to wreck the project, the narrative would need to be clear that there was evidence that that group was trying to wreck the project, rather than it being an inference of the researcher or any party who saw itself victim of that group's action. We explore the role of making inferences in Chapter 2.

How you do make sense of what happened?

The critical process with respect to articulating your sense making is making your tacit knowledge explicit. This involves providing an analysis not only of what you think is going on in the story but also of how you are making sense of it as the story unfolds. In other words, sense making is not a retrospective process only but is also a collaborative process which is concurrent with the story, and in terms of the action research cycle actually shapes the story: hence, the image we used above of the action researcher as actor-director. As you report assumptions which you held as the story progressed, you need to show how you tested them, especially if these assumptions were privately held. In terms of our example above, the researcher needs to test whether or not the group, which he thinks is out to wreck the project, actually intends that.

So what?

The third issue in action research is how the action research project is contributing theory or usable knowledge. As action research is context bound in a particular setting and set of events, it needs to have some interest and relevance to the uninvolved reader, the third person readership. Hence, the question 'So what?' is a pertinent and challenging question or, as Friedman (2001: 168) put it, 'if … then …'.

Conclusions

We have begun this chapter by seeing you poised to embark on your insider action research initiative. We have invited you to reflect on your experience of your organization so as to explore what might change through your deliberate collaborative action, involving you and others and leading to actionable knowledge that would be of use to others not directly involved.

David reflects:

> Action research expects us to stop just going through the motions, doing what we've always done because we've done it, doing it the same way because we've always done it that way. Action researchers take a close look at what they are doing and act to make things better than they already are. Taking a closer look is action in and of itself, and that research, that knowledge creation, any action taken based on that research has the potential to transform the work that we do, the working conditions that we sweat under, and most importantly the people who we are.

Enacting the action research cycle involves not only the pre-step of articulating the context and purpose of the project, and the main steps of constructing, planning action, taking action and evaluating, but also reflecting on content, process and premise issues in how the action research cycles are undertaken. Both the action research and meta learning are undertaken by individuals, by teams, between teams in the interdepartmental groups and between organizations. The rigour of your inquiry is demonstrated by how you expose these activities to critique and how your conclusions are supported by your development of theory or usable knowledge. We will now turn to how you, as the action researcher, can engage in learning in action.

Exercise 1.1   Enacting the action research cycles

Refer to Figure 1.2.

1 Select an issue/problem that you have worked on in your team (or are working on).
2 What is the *context* of this issue? Why is it important? What are the stakes involved?
3 Describe how the issue was *constructed*. How did you decide that an intervention was needed or wanted, what was wrong, what the causes were? How did you deal with different meanings or constructions in the team?
4 What action was *planned*?
5 What happened when the action was *implemented*? What were the outcomes, both intended and unintended?
6 How did the team *review* the outcomes?
7 What was *then* constructed, planned implemented etc.?
8 What is the *meta learning* from this exercise?

   a) As you look back on this, what insights do you have about the *content* of the issue? Did the initial constructing fit? Had you named the right issue? What have you learned about this issue in your business/organization?
   b) What insights do you have about *process*? How did the team work on the issue? What have you learned about how to plan, take action and evaluate?
   c) Was there any challenge to existing *premises* of how you thought about things, anything in the event that challenged the team to ask different questions, see the issue in terms of a different category of issue/problem and so on?

TWO

Learning in Action

The questions underpinning this chapter direct you to first person activities as you, the insider action researcher, engage in first and second person inquiry/ practice as you enact the action research cycles. How do you learn in action? How do you attend to what you might be learning as you engage in the issues of your action research project? How might you do research in the present? As answers to these questions, we outline the structure of human knowing and ground some processes of how adults learn in action and how reflection and journaling may be used to help you realize what and how you are learning. Our focus is not on learning *on* action but on learning *in* action.

As the insider action researcher, you are an actor in the setting of the organization. In contrast with traditional research approaches, you are not neutral but an active intervener making and helping things happen. Accordingly, a critical feature of action research is how you learn about yourself in action as you engage in first, second and third person inquiry.

At its core, first person practice means that your own beliefs, values, assumptions, ways of thinking, strategies and behaviour and so on are afforded a central place of inquiry in your action research practice – the actor-director, as we portrayed in the previous chapter. It involves attention to how you experience yourself in inquiry and in action, what Reason and Torbert (2001) refer to as 'upstream' and 'downstream' inquiry and Marshall (1999) as 'living life as inquiry'. As Marshall (2001) describes it, self-reflective practice involves enacting inquiry with intent in a manner that is distinct for each person, suggesting that each individual must craft his/her own practice and attend to its quality through inner and outer arcs of attention, enacting cycles of action and reflection and being both active and receptive.

Learning in action is grounded in the inquiry–reflection process. Schon's (1983) notion of the 'reflective practitioner' captures the essentials of knowing in action and reflection in action. Knowing in action is tacit and opens up outcomes that fall within the boundaries of what you have learned to treat as normal. Reflection in action occurs when you are in the middle of an action and you ask questions about what you are doing and what is happening around you. The outcome is immediate as it leads to an on-the-spot adjustment of your action.

Inquiry can be focused outward (e.g. what is going on in the organization, in the team etc.?) or inward (e.g. what is going on in me?). In Chapter 7 we will outline some conceptual frameworks which provide a basis for under-standing organizational processes which are utilized for that outward-focused inquiry and reflection. Here we focus on the activities of inward inquiry and reflection of first person practice. Marshall (1999; 2001) presents individual learning in action as 'inquiry as a way of being' and describes this first person research/practice in terms of (1) inquiring into the inner and outer arcs of attention, (2) engaging in cycles of action and reflection and (3) being active and receptive. Lonergan (1992; Flanagan, 1997) describes the process of grasping and internalizing the process of knowing as 'self-appropriation'.

Knowing and learning

The structure of human knowing is a three step heuristic process: experience, understanding and judgement (Lonergan, 1992; Flanagan 1997; Coghlan, 2008a; Melchin and Picard, 2008) (Table 2.1). First, we attend to our experience. Then we ask questions about our experience and receive an insight (understanding) and we follow that up by reflecting and weighing up the evidence to determine whether our insight fits the evidence or not (judgement). Of course, a great deal of our knowing is actually belief, where we accept the work of others, though at times we may need to verify for ourselves. The pattern of the three operations is invariant in that it applies to all settings of cognitional activity, whether solving a crossword clue, solving an everyday problem or engaging in scientific research. Added to the process of knowing is that of deciding and taking action.

Experiencing

Experience is the empirical level of consciousness and is an interaction of inner and outer events, or the data of sense and the data of consciousness. You can

Table 2.1   Operations of human cognition and doing

Experience	Seeing, hearing, smelling, tasting, touching, remembering, imagining, feeling, ...
Understanding	Inquiring, understanding, formulating what is being understood
Judgement	Marshalling evidence, testing, judging
Decision/action	Deliberating, valuing, deciding, choosing, taking action, behaving, ...

not only see, hear, smell, taste and touch, imagine, remember, feel and think, but also experience yourself as seeing, hearing, thinking, feeling, remembering and imagining. As the action researcher you experience a great deal as the project goes through its cycles. Some of your experiences are planned; others are unplanned. Some are what is done to you by others. Some experiences are cognitive; they occur through the intellectual processes of thinking and understanding. Some occur in feelings and emotions. At times you may feel excited, angry, frustrated, sad, lonely and so on. Other experiences may be experienced in the body: excited energy, embarrassed blushing, tightness in the stomach, headaches, ulcers or sickness. These three domains – cognitive, feelings and body awareness – are where experiencing occurs and you can learn by attending to these (Gendlin, 1981).

Understanding

Insight is an act of understanding that grasps the intelligible connections between things that previously have appeared disparate. It occurs at the intellectual level of consciousness. Sensory data are what you experience but do not yet understand. So you ask questions, 'What is this?', 'What does this mean?' Answers to such questions come in the form of insights, which are creative acts of understanding, of grasping and formulating patterns, unities, relationships and explanations in response to questions posed to your experience. The search for understanding is intelligent, focusing on a question or problem. While you might not know yet if a particular current search is intelligent, you anticipate intelligent answers. This act of understanding grasps a pattern in data. There are no recipes, rules or procedures to follow that lead inevitably to insights. The achievement of insight is unpredictable. It can happen quickly or more slowly. For example, if you do crosswords, notice your questions about a clue, the flashes of insight that you get (eventually!), how you check those insights with how they fit with the blank spaces for the letters and the other words that cross it. Then you verify: this must be the answer. Perhaps later, when you have completed other parts of the crossword, you find that you were not correct and you have another insight and then you verify that your new insight seems to fit better. If we tell you a joke, you get the unexpected connection at the end and you laugh (hopefully!). If you are watching a detective story on TV, you are furnished with clues throughout and the challenge is to figure out who the villain is with the detective as the story unfolds. You report that it suddenly dawned on you or that you saw the connection. Or it may be that you say, 'I just don't get it.' 'Getting it' occurs all the time and in all sorts of situations. Archimedes got it when, while having a bath, he saw how submerging the king's new crown would determine its specific gravity and so inform him of what it was made. You laugh when you get a joke. You get the possible answer to the crossword clue.

There are also inverse insights, ones for which there are no intelligible answers or patterns. It is not that you can't find them but that you grasp that there aren't any.

Lonergan (1992) argues that to grasp an insight into insight is to grasp knowledge about knowledge, and as such it is relevant to a whole series of basic problems in philosophy. Insight makes the difference between the tantalizing problem and the evident solution. In so far as it is the act of organizing intelligence, insight is an apprehension of relations and meaning. Every insight goes beyond experience to an explanatory organization.

Attending to experience is the first step to learning. The second step is to stand back from these experiences and inquire into them. What is it that has me feeling angry? What is it that I do not yet understand? You are reflecting on your experiences of constructing, planning action, taking action and evaluating action in the project.

Judgement

While insights are common, they are not always accurate or true. The question then is, does the insight fit the evidence? This opens up a question for reflection. Is it so? Yes or no? Maybe. I don't know. The shift in attention turns to a verification-oriented inquiry for accuracy, sureness and certainty of understanding. So you move to a new level of the cognitional process, where you marshal and weigh evidence and assess its sufficiency. You are at the rational level of consciousness. You set the judgement up conditionally; if the conditions have been fulfilled, then it must be true or accurate. There may be conflicting judgements and you may have to weigh the evidence and choose between them. If you do not think that you have sufficient evidence to assert that your insight fits the data then you can postpone judgement or make a provisional judgement and correct it later when you have other evidence.

There are, of course, such things as stupidity, obtuseness, confusion, divergent views, lack of attention and a general lack of intelligence. Understanding may not spontaneously flow from experience. Many insights may be wrong. Your interpretations of data may be superficial, inaccurate, biased. Judgements may be flawed. You can gain insight into these negative manifestations of knowing by the same threefold process of knowing.

Taking action

You are not just a knower; you also make decisions and act. Decision/action is at the responsible level of consciousness. The process of deciding is a similar process to that of knowing. You experience a situation. Using sensitivity, imagination and intelligence you seek to gain understanding as to what possible

Table 2.2 General empirical method in action research

Empirical level	Attentiveness
Intellectual level	Intelligence
Rational level	Reasonableness
Responsible level	Responsibility

courses of action there might be. At this level you ask what courses of action are open to you and you review options, weigh choices and decide. You reflect on the possible value judgements as to what is the best option, you decide to follow through the best value judgement and you take responsibility for consistency between your knowing and your doing. Accordingly, in the terms of action research, to the empirical, intellectual and rational levels of the empirical method is added the responsible level (Table 2.2).

What do you do as a result of your experience, understanding and judgement? It may be that you decide to behave differently the next time you are in a similar situation in order not to repeat the previous experience or in order to create a different outcome. What actions are you taking as a consequence of your reflection on constructing, planning action, taking action and evaluating action?

These four operations function in a cycle where experiencing, understanding, judging and deciding/acting set up another cycle of experiencing, and so on. Learning becomes a continuous cycle through life. You need to develop skills at each activity: be able to experience directly, be able to stand back and ask questions, be able to conceptualize answers to your questions, and be able to take risks and experiment in similar or new situations. The insider action research process makes particular demands on how you experience, understand, judge, decide and act.

General empirical method

The cognitional operations of experience, understanding and judgement form a general empirical method, which requires:

- attention to observable data
- envisaging possible explanations of those data
- preferring as probable or certain the explanations which provide the best account for the data.

These require the dispositions to perform the operations of attentiveness, intelligence and reasonableness, to which is added responsibility when we seek to take action.

A method is not the same as a recipe, which delivers another instance of the same product. The key to method is the relationship between questioning and answering; it is a framework for collaborative creativity that deals with different kinds of questions, each with its own objective. So questions for understanding of specific data (What is happening here?) have a different focus from questions for reflection (Does this fit?) or questions of responsibility (What ought I do?). The general empirical method of being attentive, intelligent, reasonable and responsible is a normative heuristic pattern of related and recurrent operations that yield ongoing and cumulative results. It is operative in natural science, in human sciences, in spirituality and in the common sense world of the everyday. As it is grounded, not in any thesis or grand theory, but in the recognizable operations of human inquiry and action, it crosses technical and philosophical boundaries and is applicable to inquiring from critical, constructive and constructionist perspectives. It is also the ground for first person authenticity, which we will now explore.

Authenticity

There is no guarantee that you will be attentive, intelligent, reasonable and responsible all the time. You can be inattentive and miss or ignore data. You can distort data. You can fly from insight by turning a blind eye, by refusing to ask questions, by ignoring awkward or disconfirming questions and by not facing unresolved feelings. The desire to know manifests itself in attentive questioning, but there are fears which block and divert this questioning: censoring, repressing, controlling symbols of feeling and imagining, selecting what you choose to question. You can be unreasonable in your judgements, settling for what is comfortable rather than for what the questions evoke. You can resist the evidence and try to escape responsibility. Sometimes, you might do these in ignorance; at other times you probably know that you are being bullheaded, obstinate or fearful. This is not a peculiar aberration but a frequent occurrence and you are likely to be resourceful and inventive in how you flee from attentiveness, intelligence, reasonableness and responsibility. Hence, authenticity is characterized by four process imperatives (Table 2.3). Be attentive (to the data). Be intelligent (in inquiry). Be reasonable (in making judgements). Be responsible (in making decisions and taking action) (Coghlan, 2008b). They address *process* issues in that they point to how you engage in inquiry and action and are *imperative* in that they point to what 'ought' to be. You experience data so you ought to be open to experience; hence the imperative, be attentive. So avoiding issues, closing your eyes to reality, turning a blind eye, burying your head in the sand, refusing to inquire into some matter and so on, diminish your authenticity. You ask questions and

Table 2.3   Authenticity

Operations	Activities	Process imperatives
Experience	Attending, sensing, imagining	Be attentive
Understanding	Inquiring, understanding,	Be intelligent
Judgement	Reflecting, weighing evidence, judging	Be reasonable
Decision	Deliberating, deciding, acting	Be responsible

seek answers, so you ought to question and wonder and seek to understand. Accordingly, the imperative is, be intelligent. Refusing to question or wonder, uncritically or sheepishly following the party line, suppressing curiosity and so on, destroy authenticity. You wonder if your ideas are correct, so you ought to have sound reasons for what you hold and base your judgements on evidence. So the imperative is, be reasonable. Suppressing discussion or dissent, lying about facts, obscuring evidence and so on, destroy authenticity. You discern what you ought to do, so you ought to be sensitive to value and choose what you believe to be right. The imperative, therefore, is, be responsible. Cheating, destroying resources, being unjust and so on, destroy authenticity.

Because the *core* action research project and the *thesis* research project are not identical, you need to engage your own learning in action as you participate in the action research cycles (Figure 2.1). So you are experiencing what it is like to engage in constructing, planning action, taking action and evaluating action, while inquiring and seeking insight and understanding into the enactment of the cycles, judging what is appropriate and then taking action on the basis of your judgement.

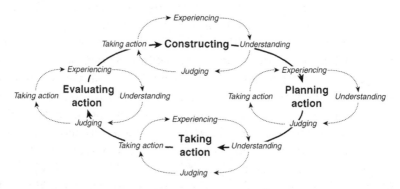

Figure 2.1   The general empirical method in action research projects

Reflection

Reflection is the process of stepping back from experience to question it and to have insights and understanding with a view to planning further action

(Kolb, 1984; Boud et al., 1985; Seibert and Daudelin, 1999; Rudolph et al., 2001; Raelin, 2008). It is the critical link between the concrete experience, the judgement and taking new action. As Raelin (2008) discusses, it is the key to learning as it enables you to develop an ability to uncover and make explicit to yourself what you have planned, discovered, and achieved in practice. He also argues that reflection must be brought into the open so that it goes beyond your privately held, taken for granted assumptions and helps you to see how your knowledge is constructed. In action research, reflection is the activity which integrates action and research. As we discussed in Chapter 1, reflection on content, process and premise is critical both to the action research cycle and to the meta learning.

Action science and developmental action inquiry, which we will introduce more formally in Chapter 3, provide focused approaches to attending to individual learning in action (Argyris, 1993; Torbert and Associates, 2004). In action science, you focus on how your actions tend to produce defensiveness and undesired outcomes, the opposite of what you intend (Argyris, 1993). This happens because you may hold assumptions which govern your behaviour, and you may make private inferences and attributions about the motives and thought processes of others which you typically do not test. Accordingly, the core of action science is learning how to identify the assumptions which govern behaviour and develop skills at testing assumptions and inferences, while at the same time exposing your own privately held theories to public testing. Through developmental action inquiry you may learn to grasp how your ability to learn in action is integrally linked to the stages of your ego development (Torbert and Associates, 2004). In the concluding sentence of their book on action science, Argyris et al. say [you are] 'iteratively moving forward from a more protective orientation toward a more reflective one' (1985: 449).

Techniques from action science, such as the ladder of inference, the right/left-hand column and treating facts as hypotheses provide valuable tools for testing consistency between the process imperatives. The ladder of inference (Figure 2.2) plots how meanings and assumptions are attributed to selected observable data and experiences, and conclusions and beliefs are adopted on which actions are based (Argyris et al., 1985; Ross, 1994). For example, at a team meeting you make a proposal for action. One of your colleagues, Joe, doesn't say anything. You think he looks as if he is sulking, and conclude that he is sulking because his proposal has not been considered. Accordingly, you decide that Joe won't be on your side and that you cannot rely on him for support, and subsequently you do not inform him of meetings as the project progresses. What has happened here is that you observed an event, i.e. all that is going on in the room. You selected part of that event (Joe not speaking) and added your own interpretations and meaning, which you did not share or test; and then your own subsequent actions of excluding Joe from

I take *actions* based on
my beliefs

I adopt *beliefs* about the person and
the situation

I draw *conclusions*

I make *assumptions* based on
the meanings added

I add *meanings* (cultural and personal)

I select *data* from what i observe

I *observe* data and experiences

Figure 2.2   The ladder of inference

further meetings were based on the beliefs and assumptions deduced from your private interpretation. In terms of the image of a ladder, you have ascended the steps of inference, from the bottom rung of what is directly observable behaviour to upper rungs of acting on privately held, untested inferences. As Fisher and Sharp (1998) express it: data are at the bottom rung, reasoning is at the middle rung and conclusions are at the top rung. Differentiating between these three elements is central to the process of learning in action. The ladder of inference helps us retrace our steps from what we have seen and heard (directly observable behaviour) to the conclusions we draw (inferences and attributions).

Argyris's (1993; 2004; Argyris et al., 1985) technique of right-hand/left-hand column provides a useful means for you to uncover your own privately held inferences and attributions in second person practice. He illustrates how that in a conversation there are privately held thoughts about the situation in the minds of the two participants, which are not shared or tested and which shape the flow of the conversation. These private thoughts are inferences and attributions and they foster defensiveness and inhibit inquiry in action. See the exercise at the end of this chapter.

Another useful construct to support learning in action is the notion of cognitive distortions, whereby you may become aware of how you might be prone to distorting reality, particularly when under pressure (Coghlan and Rashford, 1990). You may distort reality when you engage in such activities as: over-generalization, all-or-nothing thinking, mental filtering, jumping to conclusions, emotional reasoning, fortune-telling and other similar ways of misperceiving what is happening. Distortions such as these impair your ability to engage in inquiry in action.

Emotions as well as thoughts are part of the reflective process. You need to be able to recognize and acknowledge the role that feelings play in the formation of judgement and in taking action. *Focusing* (Gendlin, 1981; Cornell, 1996) provides a valuable method of listening to experiences within the body by noticing how you feel and by having a conversation with that feeling in a friendly way. It is a process of listening to your body in a gentle, accepting way and hearing the messages that your inner self is sending through your body.

In terms of the general empirical method, you are seeking to be intelligent and reasonable about what you are attending to. So you inquire of yourself what you are doing and what is going on in your head, such as: what is the evidence of your understanding? How have you come to understand in the way that you have and not in another? How do you know that your understanding fits the data? Argyris poses similar questions in a more focused way. What are your espoused theories and your theories in use, and can you express them in a way that you can't squirm out of them? How can you become more aware of your skilled incompetence, that is, how your reasoning functions to protect yourself and how you become blind to your blindness? How do you cover up inconsistent messages that you produce, deny producing them and make that denial undiscussable and the undiscussability of the undiscussable itself undiscussable? On what evidence are you forming a judgement about what is going on and what you choose to do? Is your insight an inference/attribution? How might you test it?

Developing reflective skills through journaling

Journal keeping is a significant mechanism for developing reflective skills. You note your observations and experiences in a journal, and over time learn to differentiate between different experiences and ways of dealing with them. Journal keeping helps you reflect on experiences, see how you think about them, and anticipate future experiences before you undertake them (Moon, 1999; Boud, 2001; Raelin, 2008). It enables you to integrate information and experiences which, when understood, help you understand your reasoning processes and consequent behaviour and so anticipate experiences before embarking on them. Keeping a journal regularly imposes a discipline and captures your experience of key events close to when they happen and before the passage of time changes your perception of them. McNiff et al. (2003) describe some of the useful functions a journal or research diary can have:

- a systematic and regularly kept record of events, dates and people
- an interpretive, self-evaluative account of the researcher's personal experiences, thoughts and feelings, with a view to trying to understand her own actions
- a useful way of dumping painful experiences
- a reflective account where the researcher can tease out interpretations
- an analytic tool where data can be examined and analysed.

Journals may be set to a particular structure. One obvious structure is provided in this chapter. You can keep track of your experience, the questions which arise out of the experience, the insights you receive, how you weigh evidence in order to verify your understanding and how you make decisions and what actions you take. Kolb's (1984) experiential learning cycle is a useful structure, whereby experience, reflection, conceptualization and experimentation form practical headings. These formats work well. You may learn to attend to details of a situation, and with practice can isolate critical incidents which have affected your cognitive processes and your judgement as to what to say or do. You develop skills of awareness and attentiveness. You are challenged in your use of theory, and learn to use theory in a practical manner. You learn to experience learning as a continuous life task as you apply your learning to future situations.

Another useful framework for journal keeping is Schein's (1999) ORJI model. ORJI (Observation, Reaction, Judgement, Intervention) focuses on what goes on inside your head and how it affects your covert behaviour. You observe (O), react emotionally to what you have observed (R), analyse, process and make judgements based on the observations and feelings (J) and intervene in order to make something happen (I). Schein pays particular attention to the movement from observation to judgement, because he believes that frequently the individual does not pay attention to the reaction stage. In his view, the individual typically denies feelings, short circuits them and moves straight to judgement and action. You may react to an event by saying to yourself, 'That's stupid' – a judgement. What you have probably done is to miss an emotional reaction of feeling threatened by the event. You may not have recognized or acknowledged that feeling of being threatened, yet it is present and is governing your judgement. By learning to identify and attend to feelings, (1) as initial reactions and (2) as influencing judgements, you may learn to deal with them and choose whether or not to act on them. Denial of feelings frequently means acting on them without adverting to the fact that you are acting on them. Acknowledgement of feelings to yourself and the subsequent judgement as to the origins and validity of those feelings are critical to learning and change. A journal may be structured around the four ORJI activities.

Schein's ORJI model adds a sophistication to Kolb's experiential learning cycle in two ways (Coghlan, 1993). First, it focuses on a neglected and typically bypassed area, namely the spontaneous reaction to an incident. It provides a framework whereby you may learn to recognize feelings and distinguish them from cognitive processes. Second, it inserts a structured reflection process that works back from action to judgement to reaction to observation. When your view of a situation is not confirmed by how events develop, you may question the original judgement. When you find that the judgement is based on an emotional reaction, then you may question the source of that reaction. With practice, you may learn to become more aware of emotional reactions so as to be able to recognize them as they arise, rather than in retrospect.

Second person skills

The core skills underpinning action research are relational; that is, we need to be skilled at engaging with others. When we engage in second person work, we need to be able to build relationships with others, listen well and have a range of ways of interacting with them so that collaborative inquiry and joint action can take place.

The starting point for second person practice is to examine your own dispositions. Carl Rogers (1958), in describing the characteristic of the helping relationship, invites those in the helping role to face certain questions about their own dispositions. These dispositions relate to the ability to build trust, to allow oneself to experience positive feelings towards the other, to be strong enough in oneself to allow freedom to the other, to be able to enter the subjective world of the other and to see things as he or she sees them, to be free from external evaluation, and to allow the other person to be in the process of becoming. At the same time, as we will discuss in Chapter 10, you need to be politically astute and ethical in your engagement in your second person work.

Second person research involves core skills at engaging with others in the inquiry process. In his articulation of the dynamics of helping, Schein (1999; 2009) describes several types of inquiry. His first category is what he calls *pure inquiry*. This is where you prompt the elicitation of the story of what is taking place and listen carefully and neutrally. You may ask, 'What is going on?', 'Tell me what happened.' The second type of inquiry is what Schein calls *diagnostic inquiry*, in which you begin to manage the process of how the content is analysed by the other by exploring (1) emotional processes, (2) reasoning and (3) actions. So you may ask, 'How do you feel about this?', 'Why do you think this happened?', 'What did you do?', 'What are you going to do?' and so on. The third type of inquiry is what Schein calls *confrontive inquiry*. This is where you, by sharing your own ideas, challenge the other to think from a new perspective. These ideas may refer to (1) process and (2) content. Examples of confrontive questions would be, 'Have you thought about doing this ...?', 'Have you considered that ... might be a solution?'

Schein's typology may be reframed in terms of the general empirical method: working with others to attend to their experience, to have insights into that experience, to make judgements as to whether the insights fit the evidence and then to take action (Melchin and Picard, 2008; Coghlan, 2009). Observation of people in systems and subsequent conversations between you and your co-researchers seek to bring out experience (through pure inquiry), to test insights and form judgements about that experience (through diagnostic inquiry) and then to make decisions and take action (through confrontive inquiry). Through these conversations, constructed meanings may be uncovered and tested, and action planned, taken and reviewed.

Because as insider researcher you are part of the situation, you may not always act as an external facilitator/consultant might, that is, be solely the enabler of emergent information and action. Of necessity you have a view of things as they are and what needs to change, and will be expected to share and argue that view. Accordingly, a critical skill for you as the insider action researcher is to be able to combine advocacy with inquiry, that is to present your own inferences, attributions, opinions and viewpoints as open to testing and critique (Argyris et al., 1985; Ross and Roberts, 1994). This involves illustrating inferences with relatively directly observable data and making reasoning explicit and publicly testable in the service of learning.

Argyris et al. (1985: 258–61) provide seven rules for hypothesis testing:

1   Combine advocacy with inquiry.
2   Illustrate your inferences with directly observable data.
3   Make your reasoning explicit and publicly test for agreement at each inferential step.
4   Actively seek disconfirming data and alternative explanations.
5   Affirm the making of mistakes in the service of learning.
6   Actively inquire into your own impact in the learning context.
7   Design ongoing experiments to test competing views.

Torbert and Associates (2004; Fisher et al., 2000) suggest four 'parts of speech' as useful to the action inquiry role:

• *Framing*: explicitly stating the purpose of speaking for the present occasion, that is what dilemma you are trying to resolve, sharing assumptions about the situation, and so on
• *Advocating*: explicitly stating the goal to be achieved, asserting an option, a perception, a feeling or a proposal for action
• *Illustrating*: telling a bit of the concrete story that makes the advocacy concrete and orients the others more clearly
• *Inquiring*: questioning others to understand their perspectives and views.

Putnam (1991) asks if there are recipes which might be useful in helping others explore their reasoning processes. He suggests that questions like 'What prevents you from ...?' and 'What have I said or done that leads you to believe that ...?' facilitate a focus on directly observable behaviour rather than on attribution, inference or privately held meanings. These interventions may occur in one-to-one or group situations.

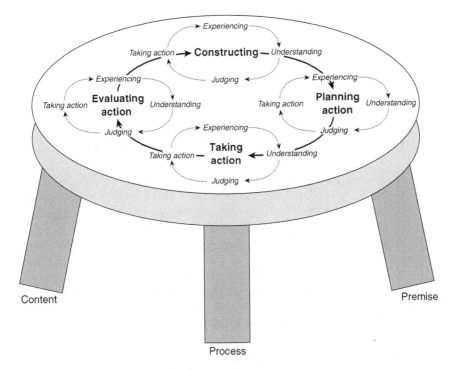

Figure 2.3   The complex dynamics of action research

Conclusions

In this chapter we have placed the focus on you as the action researcher. When you engage in the action research cycles of constructing, planning action, taking action and evaluating action with others and try to understand and shape what is going on, you are engaging in your own learning cycle activities of experiencing, understanding, judging and taking action (Figure 2.3). The general empirical method provides a normative pattern of related and recurrent operations that yield ongoing and cumulative results. It envisages all data, both of sense and of consciousness. It enables you to inquire into how you think, construct meaning, and verify your understanding as you receive insights as to what is going on in any situation and you seek to take appropriate action.

The underlying assumption is that you as the researcher are yourself an instrument in the generation of data. As Buchanan and Boddy (1992) remind us, the desire to be involved or to lead radical change involves high hassle and high vulnerability. So you need a combination of skills in self-reflection with vulnerability, realistic expectations, tolerance, humility, self-giving, self-containment and an ability to learn (Bell, 1998). When you inquire into what is going on, when you show people your train of thought and put forward hypotheses to be tested, you are generating data. Accordingly, some of your core skills are in the areas of self-awareness and sensitivity to what you think and feel within yourself and to what you observe going on around you, supported by the conceptual analytic frameworks on which you base your observations and interpretations. In this respect your knowledge base in the field of organization behaviour on which you base your observations is central.

Exercise 2.1   First person learning in action

The following exercises are aimed at stimulating your own skills in learning in action. There is an extensive literature that is aimed at helping people learn to develop skills in awareness, attentiveness and mindfulness through, for example, t'ai chi, yoga, Gestalt, Focusing and the meditation practices in all religious traditions. These are most useful. In this context we are building on these and seeking to provide tools for you to develop skills in a more focused area, namely insider action research. Fisher et al. (2000) provide very useful attention and reflective exercises that you may also practise to develop your skills in learning in action in the workplace. For illustrations of teaching reflective practice see Taylor et al. (2008).

General empirical method

- Take any puzzle with which you are confronted – crossword clue, sudoku, jigsaw, arithmetic teaser, how to get the kite down from the tree, how to prevent the water from leaking, how to help your class learn differential calculus, and so on.
- Attend to experience of movement from puzzlement or confusion to understanding through insight in your search for an intelligible solution and the flash of insight (the 'aha' moment) you receive.
- Attend to how you verify and test your insight and how you may go through many iterations of trial and error and of testing alternative insights.

Keep doing this exercise in as many of your everyday activities as is feasible or appropriate so that you learn to apprehend how you know.

Ladder of inference

Take Figure 2.2 and apply it to an incident in your project. Retrace your steps from what you saw and heard during the incident.

- What evidence did you select from all that was going on around you?
- What inferences did you draw and did not test?
- What conclusions did you draw?
- What actions did you take or not take?

Now review the whole process and receive whatever insights appear about how you may have moved from data to reasoning to conclusions.

Double column

- Take a page and write down the progress of a conversation you have with a person with whom you are working on your project.
- Then on another page or column, write down what you have been thinking privately about what is being said in the conversation and what is *not* being said.
- Notice how you make inferences and attributions privately in your own head out of what is rather hazy evidence and how you act on them by what is rather hazy evidence and how you act on them by what you say in response.

Exercise 2.2   Keeping a journal

Kolb's cycle

Based on Kolb's (1984) experiential learning cycle (McMullan and Cahoon, 1979; Coghlan, 1993).

- *Concrete experience*. Describe a concrete event which has taken place in the work situation – what happened, who said/did what, what you felt/said/did, what happened next, what the consequences were. Stick to a single event bounded by time. Be clinically neutral in the description – like a news bulletin.
- *Reflection*. Now look back with hindsight: what are your feelings, reactions, observation, judgements on this event? Perhaps now you notice that this has happened before or often. Maybe you are disappointed, angry or pleased with your own reactions at the time. How do you view your reactions and behaviour? What were the triggers that provoked your reaction?
- *Conceptualization*. Relate relevant concepts to the experience described and formulate tentative conclusions, generalizations and hypotheses.
- *Experimentation*. Suggest action implications for applying, testing and extending what you have reflected on, with a view to setting some behavioural goals for similar future situations. These are not general resolutions, but specific and concrete actions coming directly from your experience, reflection and conceptualization.

Schein's ORJI

Based on Schein's (1999; Coghlan, 1993) cycle of observation, reaction, judgement and intervention.

Take a situation or event where your own behaviour resulted in an unpredicted outcome.

(Continued)

(Continued)

- What did I actually observe? Can I describe it?
- How did I react? What feelings were aroused in me?
- What was my judgement about what happened? What thoughts or evaluations did the event trigger?
- What did I do about it? How did I intervene? (Remember that doing nothing or remaining silent is also an intervention.)

Exercise 2.3   Developing inquiry skills

In your learning group or with colleagues, form a triad and adopt roles A, B and C as follows:

- A presents an issue that she is dealing with in her action research project.
- B inquires into the issue, using Schein's (1999) intervention typology.
- C observes and then facilitates reflection on the process using Schein's intervention typology.
- Change roles and repeat.
- Change roles and repeat.

This exercise may also be done using Argyris's seven rules for hypothesis testing or Torbert's four parts of speech. McGill and Brockbank (2004) provide other useful techniques for developing skills in working in action learning sets.

THREE

Understanding Action Research

Now that you have a sense of the action and learning in action processes as you engage in your project, we step back and consider some of the theoretical issues in action research. The guiding questions underpinning this chapter are: how do I understand what I am doing in my insider action research initiative? How do I understand how it fits into a rich research tradition?

We begin by noting that understanding of the nature of research is changing. As Gibbons et al. (1994) have argued, it is time for a mode of research (which they call mode 2 research) that is transdisciplinary, heterogeneous, socially accountable, reflexive and produced in the context of application. 'The new production of knowledge', as articulated by Gibbons and his colleagues, is a network activity, and research therefore needs to follow and move away from a model whereby it is embedded currently in the expertise of isolated individuals operating from a top-down expert model (Gustavsen, 2003). MacLean et al. (2002) make the point that action-oriented research, such as action research, has the potential to meet the criteria of mode 2 research. The action orientation and collaborative nature of action research and of insider action research provide a radical alternative to the traditional doctorate (Coghlan, 2007).

Action research has been traditionally defined as an approach to research which is based on a collaborative problem solving relationship between researcher and client which aims at both solving a problem and generating new knowledge. Reason and Bradbury say that it is 'a family of practices of living inquiry ... it is not so much a methodology as an orientation to inquiry' (2008: 1). Accordingly, what is noticeable is that there is a wide diversity, not only in the practice itself, but in the discourse on action research practice. In this chapter we present some foundational perspectives on this diversity in order to help you establish some grounds for what you are doing in your insider action research project. For a flavour of discussions on this variety, you may consult such sources as Cassell and Johnson (2006), Docherty et al. (2006), *International Journal of Action Research* Special Issue (Vol. 3, Issues 1 and 2, 2007) and Reason and Bradbury (2008).

Action research as practical knowing

Action research distinguishes four kinds of knowing, reflecting different ways in which we deal with and act within the world (Heron, 1996; Reason and Torbert, 2001):

- *Experiential knowing*: the knowledge arising as we encounter the realities around us
- *Presentational knowing*: the knowledge expressed in our giving form to this experiential knowing, through language, images, music, painting and the like
- *Propositional knowing*: the knowledge distilling our experiential and presentational knowing into theories, statements and propositions
- *Practical knowing*: the knowledge that brings the other three forms of knowing to full fruition by *doing* appropriate things, skilfully and competently.

The form of knowledge that action research aims to produce is practical knowing, the knowing that shapes the quality of your moment-to-moment action.

Practical knowing has been neglected by scholars. In the seventeenth century philosophers turned to problems of the objectivity of knowing – a shift from knowing in a descriptive mode to knowing in an explanatory mode where things were no longer presented in relation to the knowing subject but were related to one another in recurring patterns. A tendency to relate any method of thinking to the subject was criticized as being subjective and invalid and limited to surface appearances, as contrasted to scientific patterns of knowing.

What we know and how we know in day-to-day living is the realm of practical knowing, or common sense knowing as Lonergan (1992) terms it. Its interests and concerns are human living and the successful performance of daily tasks and discovering immediate solutions that will work. It differs from scientific knowing in that it is particular and practical and it draws on resources of language, support of tone and volume, eloquence and facial expression, pauses, questions, omissions and so on. One of its particular characteristics is that it varies from place to place and from situation to situation. What is familiar in one place may be unfamiliar in another. What works in one setting may not work in another. What we know needs be differentiated for each situation. Understanding actions in the everyday requires inquiry into the constructions of meaning that individuals make about themselves, their situation and the world, and how their actions may be driven by assumptions and compulsions as well as by values. In a similar vein, large systems and groups hold their own shared meanings which direct their actions. Such meanings are likely to be hidden and taken for granted (Schein, 2004). Accordingly, practical knowing is always incomplete and can only be completed by attending to figuring out what is needed in situations in which one is at a given time. No two situations are identical. This is why we reason, reflect and judge in a practical pattern of knowing in order to move from one setting to another, grasping what modifications are needed and deciding how to act.

A contrast of scientific and practical knowing points to differences in how practical knowing has a concern for the practical and the particular while science has theoretical aspirations and seeks to make universal abstract statements (De Vos, 1987). Practical knowing is content with only what it needs for the moment, while scientific knowing tries to be exhaustive and seeks to know everything and state accurately and completely all it knows. Practical knowing is typically spontaneous while science is methodical. Practical knowing uses language with a range of meanings, while science develops technical jargon. In summary, practical knowing remains in the world of things related to us while scientific knowing wants to relate things to each other (Lonergan, 1992).

Foundations of action research

Action research has many origins and roots: in the work of Kurt Lewin, one of the founding fathers of social psychology, in Paolo Freire's work on consciousness-raising, in critical and pragmatic philosophy (Johansson and Lindhult, 2008), in various schools of liberation thought, notably Marxist and feminist (Brydon-Miller et al., 2003), and in Aristotelian philosophy (Eikeland, 2006a; 2008). We are building particularly on how action research developed largely from the work of Kurt Lewin and his associates, which involves a collaborative cyclical process of constructing a change situation or a problem, planning, gathering data, taking action, and then fact-finding about the results of that action in order to plan and take further action (Lewin, 1946/1997; 1948/1999; Dickens and Watkins, 1999; Coghlan and Jacobs, 2005; Bargal, 2006; Neilsen, 2006). The key idea is that action research uses a scientific approach to study the resolution of important social or organizational issues together with those who experience these issues directly.

Argyris (1993) summarizes four core themes of Lewin's work. First, Lewin integrated theory with practice by framing social science as the study of problems of real life, and he connected all problems to theory. Second, he designed research by framing the whole, and then differentiating the parts. Third, he produced constructs which could be used to generalize and understand the individual case, particularly through the researcher as intervener and his notion that one could only understand something when one tried to change it. Fourth, he was concerned with placing social science at the service of democracy, thereby changing the role of those being studied from subjects to clients so that help, if effective, could improve the quality of life and lead to more valid knowledge. Marrow, Lewin's biographer, states:

> Theory was always an intrinsic part of Lewin's search for understanding, but theory often evolved and became refined as the data unfolded, rather than being systematically detailed in advance. Lewin was led by both data and theory, each feeding the other, each guiding the research process. (1969: 128)

Argyris et al. (1985: 8–9) summarize Lewin's concept of action research as follows:

1 It involves change experiments on real problems in social systems. It focuses on a particular problem and seeks to provide assistance to the client system.
2 It, like social management more generally, involves iterative cycles of identifying a problem, planning, acting and evaluating.
3 The intended change in an action research project typically involves re-education, a term that refers to changing patterns of thinking and action that are currently well established in individuals and groups. A change intended by change agents is typically at the level of norms and values expressed in action. Effective re-education depends on participation by clients in diagnosis, fact-finding and free choice to engage in new kinds of action.
4 It challenges the status quo from a participative perspective, which is congruent with the requirements of effective re-education.
5 It is intended to contribute simultaneously to basic knowledge in social science and to social action in everyday life. High standards for developing theory and empirically testing propositions organized by theory are not to be sacrificed, nor is the relation to practice to be lost.

After Lewin's untimely death in 1947, action research became integral to the growth of the theory and practice of organization development (French and Bell, 1999; Weisbord, 2004; Burke, 2008; McArdle and Reason, 2008) and significant for organizational research (Eden and Huxham, 1996; Gummesson, 2000), such as in commercial organizations (Bradbury et al., 2008; Coughlan and Coghlan, 2009; Adler et al., 2004), education (Zeichner, 2001; Pine, 2008), community work (Stringer, 2007), health and social care (Winter and Munn-Giddings, 2001; Hughes, 2008) and nursing (Koch and Kralik, 2006).

Lippitt (1979) distinguishes three different meanings that have been denoted by the term 'action research', which reflect different roles played by the researcher. First, diagnostic research is conducted concerning some ongoing aspect of an action process. In this form of research the researcher gathers the data and presents them to those who are in a position to take some action. The research originates from the researcher's interests and is useful to the organization, partly as a payoff for allowing access. In Lippitt's view this does not constitute action research. The second meaning of the term 'action research' is connoted by a procedure of collecting data from participants in a system and providing feedback about the findings of the data as an intervention to influence, presumably in a helpful way, the ongoing action process of the system. In this model the researcher may be acting either as a data gatherer solely or in a helping role to the members of the system. The third meaning of action research is defined as a procedure in which the participants of a social system are involved in a data collection process about themselves and they utilize the data they have generated to review the facts about themselves in order to take some form of remedial or developmental action. In this model, the researcher and the researched are working in collaboration. In Lippitt's view this is the purest form of action research.

Cooperrider and Srivastva (1987) criticize how action research has developed to be viewed as a form of problem solving. They challenge what they see as underlying assumptions about the nature of action research, which are based on utilitarian and technical views of organizations as problems to be solved. As an alternative, they propose appreciative inquiry as a form of action research which focuses on building on what is already successful, rather than what is deficient.

For Gummesson, action research is 'the most demanding and far-reaching method of doing case study research' (2000: 16). He integrates the characteristics of action research from several case studies and focuses on it from a management perspective:

1 *Action researchers take action.* Action researchers are not merely observing something happening; they are actively working at making it happen.
2 *Action research always involves two goals*: solve a problem and contribute to science. As we pointed out earlier, action research is about research *in* action and does not postulate a distinction between theory and action. Hence the challenge for action researchers is both to engage in making the action happen and to stand back from the action and reflect on it as it happens in order to contribute theory to the body of knowledge.
3 *Action research is interactive.* Action research requires cooperation between the researchers and the client personnel, and continuous adjustment to new information and new events. In action research, the members of the client system are co-researchers as the action researcher is working with them on their issue so that the issue may be resolved or improved for their system and a contribution be made to the body of knowledge. As action research is a series of unfolding and unpredictable events, the actors need to work together and be able to adapt to the contingencies of the unfolding story.
4 *Action research aims at developing holistic understanding* during a project and recognizing complexity. As organizations are dynamic socio-technical systems, action researchers need to have a broad view of how the system works and be able to move between formal structural and technical and informal people subsystems. Working with organizational systems requires an ability to work with dynamic complexity, which describes how a system is complex not because of a lot of detail (detail complexity) but because of multiple causes and effects over time (Senge, 1990).

5 *Action research is fundamentally about change.* Action research is applicable to the understanding, planning and implementation of change in groups, organizations and communities. As action research is fundamentally about change, knowledge of and skill in the dynamics of organizational change are necessary. We develop this point in Chapter 5.

6 *Action research requires an understanding of the ethical framework,* values and norms within which it is used in a particular context. In action research, ethics involves authentic relationships between the action researcher and the members of the client system as to how they understand the process and take significant action. Values and norms that flow from such ethical principles typically focus on how the action researcher works with the members of the organization. We will develop this point in Chapter 10.

7 *Action research can include all types of data gathering methods.* Action research does not preclude the use of data gathering methods from traditional research. Qualitative and quantitative tools, such as interviews and surveys, are commonly used. What is important in action research is that the planning and use of these tools be well thought out with the members of the organization and be clearly integrated into the action research process. We return to this point in Chapter 7.

8 *Action research requires a breadth of preunderstanding* of the corporate or organizational environment, the conditions of business or service delivery, the structure and dynamics of operating systems and the theoretical underpinnings of such systems. Preunderstanding refers to the knowledge that the action researcher brings to the research project. Such a need for preunderstanding signals that an action research approach is inappropriate for researchers who, for example, think that all they have to do to develop grounded theory is just to go out into the field.

9 *Action research should be conducted in real time,* though retrospective action research is also acceptable. While action research is a live case study being written as it unfolds, it can also take the form of a traditional case study written in retrospect, when the written case is used as an intervention into the organization in the present. In such a situation the case performs the function of a 'learning history' and is used as an intervention to promote reflection and learning in the organization (Kleiner and Roth, 1997).

10 *The action research paradigm requires its own quality criteria.* Action research should be judged *not* by the criteria of positivist science, but rather within the criteria of its own terms.

Business consultancy language notwithstanding, Gummesson's characteristics apply to the action researcher in any organization. The research project unfolds as the cycles of planning, data gathering, taking action, reviewing and further planning and action are enacted.

Research paradigms and action research

How is action research scientific? Many writers have articulated the ontological and epistemological foundations of action research and contrasted them with those of the scientific method associated with positivistic philosophy (Susman and Evered, 1978; Riordan, 1995; Eden and Huxham, 1996; Gummesson, 2000; Reason and Torbert, 2001; Greenwood and Levin, 2007). It is not our intention to retrace those arguments here, but instead we will give a brief general overview of the three main traditions: positivism, hermeneutics and critical realism (Table 3.1).

Table 3.1   Research paradigms and action research

Philosophical foundations	Positivism	Hermeneutics and postmodernism	Critical realism and action research
Ontology	Objectivist	Subjectivist	Objectivist
Epistemology	Objectivist	Subjectivist	Subjectivist
Theory	Generalizable	Particular	Particular
Reflexivity	Methodological	Hyper	Epistemic
Role of researcher	Distanced from data	Close to data	Close to data

The philosophy of science has produced useful principles relating to epistemology and ontology which include some basic assumptions that constitute the philosophical underpinnings of warranted knowledge or theory. This in turn enables us to understand science and differing forms of explanation. Epistemology (the grounds for knowledge) and ontology (the nature of the world) can be assessed along a fairly arbitrary continuum moving from an objectivist (realist) to a subjectivist (relativist) perspectives. Researchers' epistemological and ontological perspectives legitimate their own distinctive way of doing research and determine what they consider a valid, legitimate contribution to knowledge or theory irrespective of whether we call it development, confirmation, validation, creation, building or generation (Peter and Olsen, 1993). An objectivist view of epistemology accepts the possibility of a theory-neutral language, in other words it is possible to access the external world objectively. A subjectivist view denies the possibility of a theory-neutral language. An objectivist view of ontology assumes that social and natural reality have an independent existence prior to human cognition, whereas a subjectivist ontology assumes that what we take as reality is an output of human cognitive processes (Johnson and Duberley, 2000). Different epistemological and ontological approaches encourage different kinds of reflexivity. Even though reflexivity is not a new concept to the social sciences, its importance has only come to the fore in recent times (Bourdieu, 1990).

Reflexivity is the concept used in the social sciences to explore and deal with the relationship between the researcher and the object of research. Reflection means thinking about the conditions for what one is doing, investigating the way in which the theoretical, cultural and political context of individual and intellectual involvement affects interaction with whatever is being researched, often in ways difficult to become conscious of (Alvesson and Skoldberg, 2000). Systematic reflexivity is the constant analysis of one's own theoretical and methodological presuppositions and this helps one to retain an awareness of the importance of other people's definitions and understandings (Lynch, 1999). Johnson and Duberley (2000) subdivide systematic reflexivity into two forms, methodological and epistemic. Epistemic reflexivity focuses on the researcher's belief system and is the process for analysing and challenging our meta-theoretical assumptions. Methodological reflexivity is concerned with the monitoring of our behavioural impact upon the research setting as a result of carrying out the research. This requires us to follow the research procedure and protocols identified and demanded by the different research traditions.

The dominant approach or paradigm in management and organizational studies has been positivism and its successors (explanation, hypothetico-deductive, multi-method eclecticism). These approaches are defined primarily by their view that an external reality exists and that an independent value-free researcher can examine this reality. In other words, they adhere to an objectivist (realist) ontology and an objectivist epistemology. Positivists adopt a methodological approach towards reflexivity and concentrate on improving methods and their application (Johnson and Duberley, 2000). The aim of positivist science is the creation of generalizable knowledge or covering laws. In positivist science, findings are validated by logic, measurement and the reliability achieved by the consistency of prediction and control. The positivist scientist's relationship to the setting is one of neutrality and detachment.

The hermeneutic tradition, the other main approach (sometimes referred to as phenomenological, constructivist, interpretivist, postmodern, relativist), argues that there is no objective or single knowable external reality, and that the researcher is an integral part of the research process, not separate from it. This distinction is based on the subject–object dichotomy. This ontological 'subjective versus objective' dimension concerns the assumptions that social theories make about the nature of the social world. This approach follows a subjectivist (relativist) ontology and epistemology. Inquiry is inherently value laden. Postmodernism tends to adopt a hyper-reflexivity which focuses on reflexive deconstruction of own practice. Hermeneutic inquiry is directed toward the development of particular or idiographic knowledge. Nothing can be measured without changing it and this insider perspective, close to the data, provides valid, rich and deep data.

The third approach identified by Johnson and Duberley is critical realism incorporating pragmatic action. Critical realism aligns with our concept and understanding of action research. This approach follows a subjectivist epistemology similar to the hermeneutic tradition, but an objectivist ontology like the positivists. The approach concentrates on epistemic reflexivity which looks at exposing interests and enabling emancipation through self-reflexivity.

Reflexivity is not a neutral process and is in itself socially and historically conditioned. If reflexivity is to facilitate change it needs to be guided by principles of democratic engagement and a commitment to change. Reflective knowledge has to do with normative states in social, economic and political realms. It concerns a vision of what ought to be, what is right and what is wrong, and arises through the process of consciousness-raising and conscientization (Reason and Bradbury, 2008).

The general empirical method of being attentive to data, intelligent in understanding, reasonable in judging and responsible in taking action is all-inclusive and may be the basis of the philosophical ground of inquiry in all approaches. The world is mediated by meaning which constitutes human living. We learn to construct our respective worlds by giving meaning to data that continuously impinge on us from within ourselves as well as from without. Meaning goes beyond experiencing, as what is meant is not only experienced but also something we seek to understand and to affirm. There is the task of seeking to understand the many meanings that constitute organizations and social structures, in language, in symbols and in actions (Campbell, 2000; Gergen and Gergen, 2008; Bushe and Marshak, 2008). Accordingly, we inquire into how values, behaviour and assumptions are socially constructed and embedded in meaning, and what we seek to know emerges through inquiry that attends to purposes and framing, that works actively with issues of power and multiple ways of knowing (Marshall and Reason, 2007). There is also the meaning of the world we make, through our enactment of the four territories of experience: intentions, plans, actions and outcomes (Reason and Torbert, 2001). So in terms of conscious intentionality, we enact operations of intending, planning, acting and reviewing within ourselves as first person practice, with others as second person practice, and to influence a broader impersonal audience as third person practice.

Readers undertaking an action research project through an academic dissertation will engage in their own review of these philosophical issues. Suffice it to say that action research as a scientific approach does not have to justify itself in comparison to other approaches, but rather is evaluated within its own frame of reference.

Experiential paradigms of action research

The term 'action research' is a generic one and is used to refer to a family of what might seem to be a bewildering array of activities and methods. At its core, action research is a research approach which focuses on simultaneous action and research in a collaborative manner. Within this approach are multiple paradigms or methodologies, each of which has its own distinctive emphasis (Raelin, 1999; Adler et al., 2004; Greenwood and Levin, 2007). Some action research methodologies have developed from sociology and focus on how communities as socio-political systems enact social change. These approaches have a focus outside the organizational context and tend to address structural emancipatory issues relating to, for example, education, social exclusion and power and control (Lynch, 1999; Rahman, 2008). This tradition of action research is particularly associated with action research in the southern hemisphere. Other action research approaches, particularly in the northern hemisphere, have their origins in applied behavioural science and have developed in the organizational context (Coch and French, 1948; Foster, 1972; French and Bell, 1999; Adler et al., 2004; Schein, 2008). Parallel to this approach is one that focuses on relationships, both in the workplace and between social partners in regional development (Fricke and Totterdill, 2004). The central process for building relationships is democratic dialogue. This book is addressed primarily to those working within organizational settings and as such is part of the northern hemisphere tradition of action research.

A significant feature of all action research is that the purpose of research is not simply or even primarily to contribute to the fund of knowledge in a field, or even to develop emancipatory theory, but rather to forge a more direct link between intellectual knowledge/theory and action so that each inquiry contributes directly to the flourishing of human persons and their communities (Reason and Torbert, 2001). Action research rejects the separation between thought and action which underlies the pure–applied distinction that has traditionally characterized management and social research. These approaches incorporate a collaborative enactment of action research cycles whereby the intended research outcome is the construction of actionable knowledge.

Classical action research

Action research in its traditional sense comes from the work of Kurt Lewin (1946/1997; 1948/1999) and involves a collaborative change management or problem solving relationship between researcher and client aimed at both solving a problem and generating new knowledge. The researcher and client engage in collaborative cycles of planning, taking action and evaluating. This form of action research is central to the theory and practice of organization development (Coch and French, 1948; Burnes, 2007; French and Bell, 1999). It is this form of action research that provides the central theme of this book.

Participatory action research

Participatory action research (PAR) typically has a focus outside the organizational context and involves egalitarian participation by a community to transform some aspects of its situation or structures. It focuses on concerns of power and powerlessness and how the powerless are excluded from decision making, and moves to empowering people to construct and use their own knowledge (Lykes and Mallona, 2008; Rahman, 2008). Many of the liberation or emancipatory action research approaches are variations on PAR.

Action learning

Action learning is an approach to the development of people in organizations which takes the task as the vehicle for learning. It reverses the traditional learning process where one learns something first and then applies it. In action learning the starting point is the action. It is based on two principles: 'There can be no learning without action and no (sober and deliberate) action without learning' (Revans, 1998: 83); and 'Those unable to change themselves cannot change what goes on around them' (1998: 85).

Action learning is formulated around Revans's (1998) learning formula $L = P + Q$, where L stands for learning, P for programmed learning (i.e. current knowledge in use, already known, what is in books etc.) and Q for questioning insight. Revans (1982) describes three processes central to action learning:

- A process of inquiry into the issue under consideration: its history, manifestation, what has prevented it from being resolved, what has previously been attempted. Revans calls this process system alpha.
- Action learning is science in progress through rigorous exploration of the resolution of the issue through action and reflection. He calls this system beta.
- Action learning is characterized by a quality of group interaction which enables individual critical reflection, and ultimately the learning. This is the essence of action learning and Revans calls it system gamma.

These three processes emphasize how action learning involves engagement with real issues rather than with fabrications, is scientifically rigorous in confronting the issue, and is critically subjective through managers learning in action. Participating managers take responsibility for and control of their own learning and so there is minimal use of experts (Revans, 1998; Dilworth and Willis, 2003; Pedler, 2008; Pedler and Burgoyne, 2008).

Action science

Action science is associated with the work of Chris Argyris (1993; 2004; Argyris et al., 1985; Friedman and Rogers, 2008). Argyris places an emphasis on the cognitive processes of individuals' 'theories in use', which he describes in terms of model I (strategies of control, self-protection, defensiveness and covering up embarrassment) and model II (strategies eliciting valid information, free choice and commitment). Attention to how individuals' theories in use create organizational defensiveness is an important approach to organizational learning (Argyris and Schon, 1996; Senge, 1990).

Developmental action inquiry

Developmental action inquiry is associated with the work of Bill Torbert (2004; Fisher et al., 2000; Torbert and Taylor, 2008). Torbert defines action inquiry as 'a kind of scientific inquiry that is conducted in everyday life … that deals primarily with "primary" data encountered "on-line" in the midst of perception and action' (1991: 220). Torbert adds the developmental dynamic of learning to inquiry in action, emphasizing that as we progress through adulthood we may intentionally develop new 'action logics' through stages of development. Developmental theory offers an understanding of our transformation through a series of stages so that we gain insight into our own action logic as we take action.

Cooperative inquiry

One of the forms that action research takes is cooperative inquiry (Reason, 1988; 1999; Heron, 1996; Heron and Reason, 2008). Heron and Reason define cooperative inquiry 'as a form of second person action research in which all participants work together in an inquiry group as co-researchers and co-subjects' (2008: 366). Reason (1999) set out the process of cooperative inquiry in the following stages:

1. The group talks about the group's interests and concerns, agrees on the focus of the inquiry, and develops together a set of questions or proposals its members wish to explore.
2. The group applies actions in the everyday work of the members, who initiate the actions and observe and record the outcomes of their own and each other's behaviour.
3. The group members as co-researchers become fully immersed in their experience. They may deepen into the experience or they may be led away from the original ideas and proposals into new fields, unpredicted action and creative insights.
4. After an agreed period engaged in phases 2 and 3, the co-researchers reassemble to consider their original questions in the light of their experience.

In Chapter 8 we present a case example where Meehan used cooperative inquiry in his own organization.

Clinical inquiry/research

In writing about an organization development approach to organizational research, Schein (1995; 2008) introduces the notion of the 'clinical' approach to research. For Schein, 'clinical' refers to those trained helpers (such as clinical and counselling psychologists, social workers, organization development consultants) who work professionally with human systems. These trained helpers act as organizational clinicians in that they (1) emphasize in-depth observation of learning and change processes, (2) emphasize the effects of interventions, (3) operate from models of what it is to function as a healthy system and focus on pathologies, puzzles and anomalies which illustrate deviations from healthy functioning, and (4) build theory and empirical knowledge through developing concepts which capture the real dynamics of systems (Schein, 1997). At the heart of clinical inquiry is the engagement of the consultant with clients to help clients perceive, understand and act on events by helping them explore their experiences and reflect on their insights about these experiences so as to come to judgements and make decisions and take action (Coghlan, 2009).

Appreciative inquiry

Appreciative inquiry has emerged from the work of Cooperrider, and aims at large system change through an appreciative focus on what already works in a system, rather than a focus on what is deficient (Cooperrider and Whitney, 2005; Reed, 2007; Ludema and Fry, 2008). It is built around four phases:

1 Discovery: appreciating the best of 'what is'
2 Dream: envisioning 'what could be'
3 Design: co-constructing 'what should be'
4 Destiny: sustaining 'what will be'.

Appreciative inquiry takes a counter view to clinical inquiry, through its focus on appreciation rather than pathologies and problems.

Learning history

A learning history is a document composed by participants in a change effort, with the help of external consultants who act as 'learning historians' (Kleiner and Roth, 1997; Roth and Bradbury, 2008). It presents the experiences and understandings in their own words of those who have gone through and/or been affected by the change in order to help the organization move forward. The learning history is an action research process by being an intervention into the organization. This happens when the action research documentation is made available to organizational stakeholders as 'a written narrative of a company's recent set of critical episodes' (Kleiner and Roth, 1997: 173) with the purpose of facilitating learning. Kleiner and Roth (1997) present a framework for how this might be done. The narrative is read by significant stakeholders who contribute to the story from their perspective in a special right-hand column on the page. Those social scientists and 'learning historians' who study the narrative use a left-hand column for their reflection and analysis as the basis for further discussion in the organization.

Collaborative management research

Collaborative management research

> is an effort by one or more parties, at least one of which is a member of an organization and at least one of whom is an external researcher, to work together in learning about how the behaviour of managers, management methods or organizational arrangements affect outcomes in the system or systems under study, using methods that are scientifically based and intended to reduce the likelihood of drawing false conclusions from the data collected, with the intent of improving performance of the system and adding to the broader body of knowledge in the field of management. (Shani et al., 2008: 20)

Shani et al. (2008) understand that collaborative management research is unique and different from action research in that it seeks to add value to the action research approaches discussed in this chapter through how practitioners and researchers engage in a joint undertaking where each partner takes some responsibility for the others partners' learning and knowledge.

There are other expressions of action research that stress some particular emphases. *Intervention research* is a term that expresses the form that some action research takes in France in how researchers work with companies (Hatchuel and David, 2008; Buono and Savall, 2007). *Interactive research* is a term that is used in the Nordic countries. This action research approach emphasizes the relationships between participants as among equals and a high degree of participation. It stresses the joint learning that goes on between participants and researchers throughout the entire research process (Nielsen and Svensson, 2006). *Synergic inquiry* is a form of collaborative inquiry that has its root in Chinese philosophy (Tang and Joiner, 2006).

Reflective practice

Reflective practice refers to how individuals engage in critical reflection on their own action. It is associated with the work of Schon (1983; 1987). Reflective practice may be viewed as a specific dimension of action research, i.e. first person inquiry. By and large, published accounts of reflective practice focus only on the individual and generally do not consider any organizational dynamics or outcomes related to the individual's action. Schon (1983) reflects on four ways that reflective practitioners might engage in 'reflective research':

- *Frame analysis*: when practitioners become aware of their 'frames' and consider alternatives
- *Repertoire building research*: accumulating and describing examples of reflection in action
- *Research on fundamental methods of inquiry and overarching theories*: by examining episodes of practice in an action science
- *Research on the process of reflection in action*: studying processes whereby practitioners learn to reflect in action.

Evaluative inquiry

Closely related to action research is the process of evaluative inquiry, which is a reformulation of traditional evaluation practices through an emphasis on using the process of inquiry to generate organizational learning (Preskill and Torres, 1999). Many of the processes within action research, such as collaborative inquiry, reflection, joint planning and taking action, are utilized as interventions to shape how projects are evaluated in order to stimulate organizational learning.

For the neophyte reader, these multiple methodologies may be confusing. In our view, it is important to emphasize that these different methodologies are not mutually exclusive. They are sets of general principles and devices which can be adapted to different research issues and contexts. Each has its own emphasis and can be appropriately used in conjunction with other approaches. What is important is that you, as the action researcher, be helped to seek the method appropriate to your inquiry and situation.

Conclusions

In this chapter we have stepped back from the action dynamic of action research in order to explore the nature of action research and to assist you to ground your work in the rich and deep tradition of action research. We have outlined the foundations of action research as research which is based on a collaborative action relationship between researcher and members of an organization or community which aims both at addressing an issue or solving a problem and at generating new knowledge. Regretfully, it has often become a glib term for involving clients in research and has lost its role as a powerful conceptual tool for uncovering truth on which action can be taken. Action research is a form of science in the field of practical knowing, which differs from the model of experimental physics, but is genuinely scientific in the emphasis on careful observation and study of the effects of the behaviour on human systems as their members manage change.

> Questions of method are secondary to questions of paradigm, which we define as the basic belief system or worldview that guides the investigator, not only in choice of method but in ontologically and epistemologically fundamental ways. (Guba and Lincoln, 1994: 105)

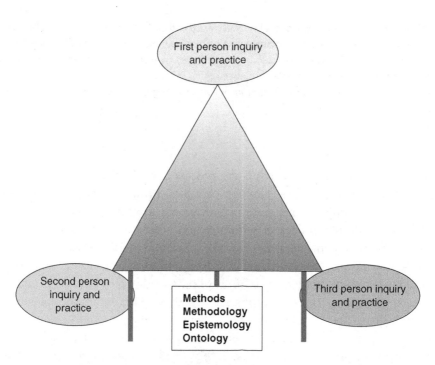

Figure 3.1 Integrated perpective on action research

Based on the point being made by Guba and Lincoln, Figure 3.1 provides a visual framework for gathering together the themes of Chapters 1, 2 and 3 for your insider action research initiative. The processes of action research through first, second and third person inquiry and practice stand on the pillars of ontology, epistemology, methodology and methods. At the base is your world-view, i.e. your ontology, which reflects your understanding of yourself, your own experience, the nature of the relational world and the nature of knowledge and theory. Epistemology (theory of knowledge) expresses how you seek to know. Methodology articulates the approach you are adopting to inquiry. Finally ontology, epistemology and methodology find concrete expression in the design and implementation of your methods for attending to, understanding, judging and enacting the collaborative activities and for framing your theory in your insider action research initiative.

Part II

Implementation

FOUR

Framing and Selecting Your Insider Project

The question underpinning this chapter is: how might you might frame and select an action research project? When you 'frame' an issue you are naming it, and by naming it you are focusing on how you might set up its analysis and set the criteria for evaluation. Framing is a heuristic process, by which we mean that the definition of an issue already includes elements of the solution. 'Reframing' is the process whereby you question an existing frame and possibly discard it in favour of a different one. Then, new frames of reference need to be created to reframe the issues in such a way that problem solving can be effective.

Framing the action research project

Framing an issue can be a complex process. It may be that what is attractive to you as an insider action research project is a practical operational issue; there is a recurring problem, which management or superiors would like researched and solved. Bartunek et al. (2000) present an example of where a project was identified as one of improving relationships between the bank and a client. Such a research project can clearly meet criteria of being useful, particularly to management, achievable in the research time frame and manageable for the researcher.

It may also turn out as the research progresses that this apparent operational problem is more complex than it appears, requiring key people to alter their mandates and ways of thinking. Bartunek et al. (2000) demonstrate how, in a manufacturing plant, the development of an integrated manufacturing system involved radical changes in how the company did business.

The complexity of issue identification and selection illustrates that the search for an appropriate issue to study is difficult. How do you get a sense of the array of possible issues which may be addressed? By using the term 'array' we are acknowledging the existence of a wide and diverse set of issues all vying for research attention. That is not to say of course that all issues are immediately apparent to you. Some may be blatantly obvious while others may go unnoticed unless attempts are made to uncover organizational members' perceptions of key issues. Not every issue will volunteer itself automatically for resolution. It is human perception that makes the difference, thus leading us to conclude that organizational actors' interpretations are pivotal in this whole process.

While acknowledging the existence of a wide and diverse array of issues, it is important to understand that any issue once selected for attention may be embedded in a set of related issues (Beckhard and Harris, 1987). You are then confronted with choices concerning boundaries and are obliged to choose between what can be achieved within the time specified for your research and available resources.

Thinking in terms of issues, rather than problems or opportunities, which warrant attention is vital, as language and labels are of the utmost importance at the outset (Dutton et al., 1983; Cooperrider and Srivastva, 1987). For example, framing proposed research initiatives in the context of addressing problems or opportunities carries some inherent risks. Framing an issue as a problem may influence who gets involved in problem resolution. It may be that organizational members embrace problems with a sense of loss, wondering about the organization's ability to reach a satisfactory resolution and often preferring to remain somewhat detached and uncommitted. The action research project may be challenging traditional procedures and ways of thinking.

Using the word 'problem' as distinct from 'opportunity' may also lead to convergent thinking (Dutton et al., 1983). The mental effort expended on problem resolution may restrict the range of alternatives considered, blinding organizational members to the possible existence of novel solutions. In a similar vein the use of the label 'opportunity' may lead to divergent thinking as this label has a greater sense of gain associated with it. Organizational members may feel a sense of excitement about tackling a significant opportunity which may have the potential for creativity.

Finally, language and labels are important as they have the potential to influence risk-taking behaviour (Dutton et al., 1983). It may be that thinking in terms of opportunities cultivates a risk-taking culture, while thinking in terms of problems cultivates a risk-averse culture. If you think with an opportunity mindset then you are less likely to embark on a witch-hunt looking for someone to blame, as there isn't anything for which to blame them, while the mindset associated with problems embraces the notion of finding a scapegoat. It seems obvious from the above that there is merit in thinking in terms of issues without any attempt to subclassify such issues in the first instance.

Williander and Styhre (2006) report on how Williander designed a collaborative research proposal to serve as an intervention to develop an individual experience among corporate engineers, product planners and other stakeholders. A later project stage would focus on the dissemination of that individual experience in order to develop an organizational capability at Volvo regarding how to capitalize on the common good. It was assumed that an intervention research approach might positively affect the time required for the company to change and build capability, since it might avoid using a trial-and-error approach, span the area from individual via organizational to institutional change, and benefit from the multidisciplinary research setting and the rigour of the research methodology.

What becomes important then is to uncover the issues which are viewed by organizational members as key issues warranting attention at any point in time. In those cases which involve complex organizational change, many of these key issues may initially fall in the category of operational problem solving. As already noted, not all issues are blatantly obvious and it is therefore important for you as the researcher to get a sense of both the obvious and the less obvious. It may be that the obvious is but an outward manifestation of a deeper issue which organizational members are not so willing to embrace publicly. Identification of these deeper issues may point to the need for inquiry into the fundamental assumptions which keep a problem recurring. What if the obvious only seems that way due to people being ill-informed on the nature of the issue at hand? Could it be that the obvious has become so as it embraces the language of dissent and reflects the preoccupation of organizational members with consequences without them ever reflecting on root causes?

Williander and Styhre (2006) reflect that when the research proposal was brought to Volvo's senior management and approved, there was no budget reserved for it. Even though the insider action researcher received strong support for his project, at least in the form of lip service, there was no money available that could be released to finance it in the current fiscal year.

As insider action researcher you need to go with the story as it evolves. As the initial questions and data demonstrate that they are inadequate for addressing the issues, you work at keeping inquiry active. You are continually testing as to whether consensus exists concerning the array of issues which could be addressed. Such an array may be constructed having considered organizational members' perceptions of key issues. It may embrace a healthy diversity of thinking among organizational members, or alternatively it may point to significant pockets of conflict in certain issue domains. Change triggers discussion, debate and arguments between people who champion competing ideas and proposals. Such discussion provides useful data and is desirable in order to expose different ideas to public scrutiny and examination.

Regardless of what the array of issues reflects, it is impossible to construct research without leaving one's own familiar world and entering into the world of others through open and honest dialogue. That of necessity means that you need to be willing to explore key concepts and themes and attempt to construct the perceptions of others concerning the range of issues. It involves understanding why organizational members frame such issues in the first instance, while simultaneously capturing causal relationships (Dutton et al., 1983).

No issue in an organization is context free (Dutton and Ottensmeyer, 1987). Uncovering issues necessitates not only establishing multiple versions of 'the real facts' but also understanding the role that history and experience have to play in organizational members' perceptions of these facts. In a similar manner, any given issue may be embedded in a system of political behaviour which is critical to understand if issue resolution is ever to be negotiated.

Moore (2007), in his inquiry into how he, as chief executive of a charity organization, might improve governance practice and improvement, found parallels between his experience of undertaking insider action research and the biblical image of the original sin committed by Adam and Eve when they ate the forbidden fruit from the tree of knowledge. Accordingly, he traces his experience in those allegorical terms – feeling he was in the Garden of Eden at the start of his research and finding forbidden fruits as he descended into the details of the norms of the organization through his questioning. He describes his sin as bringing to light and starting to taste his underlying assumptions which challenged the dominant views and values that held sway in the organization's board. Then, he found that he had exposed perceptions that he had not previously encountered and was faced with an inherent conflict between his roles as an insider and as a researcher, an experience to which he refers as realizing that he was naked. He concludes that having eaten of the tree of knowledge he found the tension of trying to occupy the worlds of both knowledge and innocence unbearable – and he resigned.

You might read Moore's (2007) article in full to get the richer picture of his understanding of his insider action research experience.

Questions for reflection and discussion

- How do you relate to Moore's image of the Garden of Eden for his insider action research project?
- How might an image help frame an action research initiative? What insights might flow from such an image?
- Is there any image emerging for you for your insider action research project?

The process of identifying issues may be characterized as fluid, dynamic and emergent (Dutton et al., 1983). It is fluid in the sense that it is difficult to establish precise boundaries, and when such boundaries are established they are often subject to change. It is dynamic in the sense that the core focus is subject to continuous revision as understanding deepens. It is emergent in the sense that issues appear over time. These key characteristics point to a process which is further characterized by the unfolding nature of interpretation and reinterpretation, making extensive use of organizational members' judgements and revision of judgements based on insights gained from new and existing data, stimuli and perceptions.

Of immediate importance then, to you, is the need to gather and organize these data, stimuli and perceptions of yourself and others. The subsequent sense making process points to the need for you to have good organizational and analytical capabilities. Krim (1988) kept a journal of his reflections and observations, and used his academic supervisors to test them in a safe environment.

In the context of deriving meaning, it is useful at this juncture in the research process to frame issues in broad categories without attempting to attach a single dominant interpretation to any issue. The existence of multiple interpretations concerning an issue must not be eliminated. Capturing multiple and diverse interpretations adds to a deeper, richer picture of the issue at hand and holds the key to more effective resolution for the long term.

There is an inherent danger in the process of attempting to simplify an issue by reducing ambiguity at an early stage. Such endeavours are manifested by ignoring some interpretations of an issue and attempting to attach a single dominant interpretation with a view to aiding resolution. Such a process may seem quite rational to the researcher who may be eager to get on with the task at hand. Rationality plays a role in the analysis stage but one cannot assume that rational analysis will lead to resolution. Resolution often involves a process of negotiation, embracing a sense of give and take where political interests warrant careful management. Neglecting political influences is a recipe for inaction as any proposed course of action may be planned to death and eventually be stillborn.

Attaching a dominant interpretation to an issue is not necessarily bad as long as you remain cognizant of the fact that other interpretations exist and are willing to test such interpretations as the need arises. It is equally important to understand the basis for such dominance. Does the interpretation reflect a shared mind at all or almost all organizational levels, or does it reflect the shared mind of a particular group such as management or trade unions?

The extent to which a dominant interpretation of an event or issue is shared or otherwise by organizational members is important for you as the researcher, as it implies that different strategies must be employed to aid issue resolution. Where a dominant interpretation is widely shared you are more likely to gain a greater degree of commitment to the resolution process with lower levels of political activity, at least at the early stages. Where the interpretation reflects the mind of a specific group, you need to be cognizant of the fact that other groups may not share that interpretation and may choose to never share it for political reasons. Friedman (2001) presents projects initiated by individuals on a continuum, with technical issues at one extreme and non-technical at the other. In his view, technical operational issues are easier to gain support for, while non-technical issues are harder to influence and are embedded in organizational defensive routines.

Finally, it is important when categorizing issues that each issue is framed in the context of its implicit and explicit assumptions, any known causal relationships, and any predictive judgements concerning the speed of issue resolution (Dutton et al., 1983). Making assumptions explicit aids the resolution process as organizational members develop a shared understanding of the issue being addressed in terms of its history, scope and possible outcomes. Establishing causal relationships helps to place an issue in context by grounding it in organizational reality while simultaneously establishing how organizational members attribute certain outcomes to root causes. Outlining predictive judgements attaches a sense of urgency or otherwise to the issue at hand.

Angela, working in a professional public health service, sought to devote her insider action research project to the problem of client waiting lists and to work with her colleagues on reducing the time clients had to wait for an appointment and for treatment. With such a clear framing of the initiative on what was accepted as a pressing problem, Angela received widespread support from her colleagues and management. She recruited her colleagues to engage in a cooperative inquiry process to explore the problem of the waiting lists and to engage in action research to resolve it as best they could. As the cooperative group met over several weeks, Angela found the group was showing a high dependence on her, e.g. waiting for her to chair the meetings and to keep the initiative alive and moving forward. While she had insights that this was a normal pattern to be expected in a developing group (Wheelan, 1999), her generative insight was that this whole process of jointly exploring a shared concern was radically new for her colleagues. Discussion of the problem of waiting lists led to discussions on priorities and the values underpinning the service and it was apparent that this group had never engaged in such conversations. In short, this was a professional organization in which each member had an individual caseload and there was no tradition of the members meeting as a group to explore core values, priorities and strategies. It was this agenda that began to dominate the group meetings. Accordingly, as Angela was confined to a particular schedule in which to deliver her master's dissertation and it was clear that the topic of waiting lists would not now be addressed adequately within that time frame, she reframed her dissertation topic to focus more on how individual professionals formed a more cohesive group. The issue of waiting lists remained a concern for the service and would be addressed consequently at a later date, but would not be the subject of Angela's dissertation.

Questions for reflection and discussion

- How does Angela's reframing of her dissertation topic make sense?
- How are you currently framing your insider action research initiative? Might that change?
- What might make it change? How do you keep your options open?

Selecting the research project

Having identified a range of issues, you are confronted with selecting an issue or issues for immediate attention in the context of a specific research agenda. Before making a final selection you are well advised to reflect on each issue identified from personal and organizational perspectives with a view to establishing:

1 The degree to which it offers an opportunity to experiment with existing and/or newly acquired knowledge
2 The degree to which it offers opportunities for personal growth and learning
3 The degree to which issue resolution offers the possibility of increasing your profile within the organization
4 The balance between personal gain and organizational gain in the event of successful resolution
5 The degree to which the issue may be resolved within known resource and time constraints.

Bjorkman and Sundgren (2005) reflect on their respective insider action research projects and note that obtaining management attention, creating interest among colleagues, creating and maintaining legitimacy, and allocating time and financial resources are key strategies. They draw their shared experience together and suggest four features of framing and selecting an insider action research initiative:

- finding 'red and hot' issues for the organization
- positioning the relational platform
- using the inside
- using and diffusing results.

Questions for reflection and discussion

- Do you know of a 'red and hot' issue for your organization that you might and are in a position to take up?
- How might you position your 'relational platform'?
- What are your strengths as an insider to take on this project and how might you use them?

Writing an insider action research proposal

The dissertation proposal is your statement that what you intend to do is such that you know what you are getting yourself into (through showing familiarity with both the organizational setting and some relevant literature), that what you intend to do is worthwhile, and that you have a potential contribution to knowledge to make. You are making it to a specific audience, the school's research application committee, your doctoral committee or some similar body.

There are four areas that your insider action research proposal needs to address: context, action, research and the insider process.

- *Context* here refers to the social and academic context of the research. There are three context areas: (1) broad general context at global and national level; (2) local organizational/ discipline context, that is what is going on in your organization; and then (3) your specific topic area. In action research, social context is very important. Therefore, you need to describe your own organization or community. This would include details of the organization, community or group, what it does, some historical background, what its concerns are, what the issues in which you are engaging with it mean, and what is intended and hoped for out of the action research project. This description contains not only a presentation of the facts of the organization in its setting, but also an introductory literature on the setting. For instance, participants undertaking action research in the master's programme in health service policy and management at Trinity College, Dublin typically have a chapter describing general developments in the Irish health system, locating their action research in national strategy and then in developments in their particular discipline or service. The thesis proposal would outline some of the basics of the context in order to show that you know what you are getting yourself into.
- *Action* covers the basic thrust of the action you intend to take or to lead. The headings in the process of change section in Chapter 5 provide an important construct in this regard. What is the action? What is the rationale for this action? Why is it worth doing? What is the desired future? What is the present situation? What is the plan to move from here to there? What is the time schedule? With whom will you collaborate? Where do you (as the researcher) fit into the action? What are the ethical challenges?
- *Research* then describes the focus of inquiry in action. What is the rationale for researching this action? What is the contribution to knowledge that this research intends to make? How do you intend to inquire into the action? How do you ensure quality and rigour in your action research?
- *Insider action research* brings the focus specifically on insider inquiry. How do you do action research in your own organization? How will you manage preunderstanding, role duality and organizational politics? (These topics are discussed in Chapters 8, 9 and 10.) How will you engage others? What are the ethical issues, particularly if you have to apply for ethical approval from a review board? (See the discussion in Chapter 10.)

Remember that action research is a different form of research from traditional research:

- traditional research begins with what we know and seeks to find what we don't know
- action research begins with what we don't know and seeks to find what we don't know
- what we don't know that we don't know is the particular fruit of action research.

As we discussed in Chapter 1, there are two action research projects running concurrently, what Zuber-Skerritt and Perry (2002) call the *core* action research project and the *thesis* action research project. Now we are focusing on writing the thesis action research proposal, which is the inquiry in action into how the core action research project will be designed, implemented and evaluated and how you intend to enact your role in it and how you intend to reflect on it.

Developing the action research

As you embark on your project, you will be moving from the preliminary definition of the issue to developing a deep, broad understanding of the issue and its context. You will begin to identify pressures in the system (and in the larger context) which affect the situation. You will be learning to identify key individuals and groups whose support is required to bring about change. Who has knowledge and understanding? Whose support is critical to developing and conducting the action research and its outcomes? As you enact cycles of action and reflection, you will discover new information. New constraints will emerge.

In parallel, you will need to be working on building collaborative relationships: contracting, building rapport, negotiating roles and levels of involvement. If you have pre-existing patterns of relationships the challenges of role duality may have an impact.

As part of her insider action research project for her master's, Margaret was engaging her own team in a collaborative inquiry process on how to improve the occupational therapy service in which they were engaged. A particular challenge for her was that her team, particularly in the initial stages, saw themselves as helping her on her action research project. Accordingly, they were deferential to her with regard to what they would explore at the meetings and at one point cancelled a meeting because she would not be present: in the words of one of the members, 'There would be no point if you're not there.' Margaret, therefore, had to put a great deal of effort into getting the group to own the project of service improvement themselves and so to engage in collaborative inquiry. She did this by rotating the chairing and note-taking role.

Questions for reflection and discussion

- Does this story evoke any resonances with your project?
- How would you deal with the dilemma of dependence?

Conclusions

Framing and selecting an action research project is a complex matter. What appears clear at the outset may lose its apparent clarity as the project unfolds. How you frame and subsequently reframe the project may hold important learning for you. The critical issue for you is to be able to frame and select a project from a position of being close to the issue. The acts of framing and selecting your action research project are themselves action research learning cycles. In other words, you do your initial framing, reflect on how that framing fits or not, articulate some understanding of why that framing fits or doesn't fit, and then take action accordingly and so test that situation. Similarly, you make your initial selection, test it and adapt it according to the data generated by the selection and framing processes.

Exercise 4.1   Questions for framing and selecting

As you look your organization/section in which you work:

1   What questions arise out of your experience to which you would like to search for answers?
2   What might be the answers to these questions?
3   What do you think might be the underlying causes of the situation for which you have these questions?
4   What alternative answers might exist?
5   Where do you fit into the situation as defined by the questions?
6   What would other members of the organization think of you working on this issue?
7   What support will you have?
8   What opposition will you encounter?
9   Where are the sensitive issues?
10   What are the constraints?
11   Who needs to be involved? Whose support do you enlist?
12   Where would be a good place to begin?
13   How will you engage in uncovering the data?

Exercise 4.2   Writing a thesis proposal

Context
What are the current issues and trends in your area of focus?

1   The broad external and internal context of the environment in which your organization is located. Examples are what's going on, the strategy policy documents etc.
2   The local organizational/discipline context. How is your organization situated in relation to the broader context? What is going on in your hospital, school, business, service?
3   Your particular topic issue. Where does that fit into the above two contexts? What will your work contribute in these contexts?

Action

- What is the action?
- What is the rationale for this action?
- Why is it worth doing?
- What is the desired future?

- What is the present situation?
- What is the plan to move from here to there?
- What is the time schedule?
- Where do you (as the researcher) fit into the action?
- What are the ethical challenges both within the project itself and in seeking approval from an ethics review board?

Research

- What is the rationale for researching this action?
- What is the contribution to knowledge that this research intends to make?
- Why should anyone not directly involved be interested in this?
- How do you intend to inquire into the action?
- How do you ensure quality and rigour in your action research?

Insider action research

- How do you do action research in your own organization?
- How will you manage preunderstanding, role duality and organizational politics?
- How do you plan to learn in action?
- What quality processes will you build into your initiative?
- What are the particular ethical challanges that you will face?

FIVE

Designing and Implementing Your Action Research Project

The questions underpinning this chapter are: how do you design your insider action research project? How do you implement it? In this chapter we explore how you may design and implement an action research project in an organization, be it a company, a hospital, a school or a unit/department or service within the organization. The action research process in your own organization follows the action research cycle introduced in Chapter 1, which involves:

1 Constructing the initiative with relevant stakeholders and systematically generating and collecting research data about an ongoing system relative to some objective or need
2 Engaging with others in reviewing the data generated and collected
3 Conducting a collaborative analysis of the data
4 Planning and taking collaborative action based on shared inquiry
5 Jointly evaluating the results of that action, leading to further planning.

So the cycle is repeated.

The subject of organizational change is a wide and complex one with an extensive literature. Mitki et al. (2000) cluster change programmes as limited, focused and holistic:

- *Limited change programmes* are aimed at addressing a specific problem, such as team building or communication improvement. It is likely that such projects may be appropriate outlets for insider action research for MBA or other master's students.
- *Focused change programmes* identify a few key aspects, such as time, quality or customer value, and then use these by design as levers for changing the organization system-wide. This type is also likely to be attractive for insider action research for MBA and other master's programmes.
- *Holistic change programmes* are aimed by design to simultaneously address all (or most) aspects of the organization.

The key question you need to ask at the outset is what type of change is at stake in your insider action research initiative. As indicated above, limited and focused change programmes fit well with limited-term master's dissertation contexts where you have a specific focus for your research that needs to be completed in a year or thereabouts. Holistic change, of its nature, takes more time and is likely to be the focus for a doctorate.

Buono and Kerber (2008) describe three approaches to change.

- *Directed change* is where there are tightly defined goals and the leadership directs and commands
- *Planned change* is where there is a clear goal and vision of the future and the leadership devises a roadmap to reach it and influences how it is reached
- *Guiding change* is where the direction is loosely defined and the leadership points the way and keeps watch over the process. This approach to change fits with much of what is written about the 'new science' of leadership and draws on constructivism and postmodernism as a foundation for dialogue and conversation about change (Bushe and Marshak, 2008).

Again you need to consider your approach to change in your insider action research project. If you are the agent of a tightly controlled and directed change approach, then you may find that it conflicts with action research values of participation. The planned change approach provides a very useful format for most insider action research projects and the process for such an approach is presented later in this chapter. Scope for the guiding change approach is more likely to be linked to action research at doctoral level.

The process of planned change

Taking planned change as the most likely approach for insider action research, especially at master's level, the question then is: how do you go about implementing the action research cycle in a planned way in a large system? While actor-directors go with the story in their film making, they also create and follow a script. The process whereby the action research agenda is identified and worked through has been well articulated by Richard Beckhard (Beckhard and Harris, 1987; Beckhard and Pritchard, 1992; Nadler, 1998; Coghlan and Rashford, 2006). Beckhard's framework has four phases (Figure 5.1):

1 Determining the need for change
2 Defining the future state
3 Assessing the present in terms of the future to determine the work to be done
4 Managing the transition.

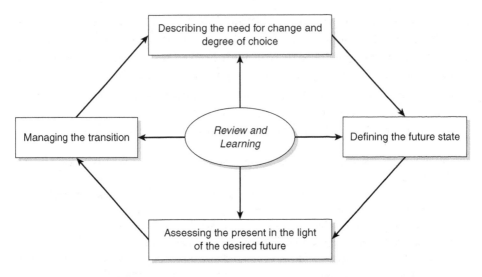

Figure 5.1   The process of change

As we will discuss in Chapters 9 and 10, doing action research in your own organization is intensely political and involves you in concurrent and sometimes conflicting roles. We think that it is important to remind you that managing the political system at every step is more important than any rigid adherence to an idealized picture of how these steps might work.

Determining the need for change

The preferred starting place is to inquire into the context for change in the organization, unit or subunit. It may seem obvious that naming the need for change and its causes is essential. The forces for change may be coming from the external environment, such as global market demands, developing customer needs and so on. They may be coming from the internal environment, such as budget overruns, low morale among staff, excessive dysfunctional political intergroup rivalry and so on. Understanding these forces identifies their source, their potency and the nature of the demands they are making on the system. These forces for change have to be assessed so that major change forces are distinguished from the minor ones.

A second key element in evaluating the need for change is the degree of choice about whether to change or not. This is often an overlooked question. Choices are not absolute. While there may be no control over the forces demanding change, there is likely to be a great deal of control over how to respond to those forces. In that case there is likely to be a good deal of scope as to what changes, how, and in what time scale the change can take place. The action research cycle enables shared inquiry into how these forces for change are having an impact and what choices exists to confront them. The outcome of determining the need for change is to ask a further question, which is whether first or second order change is required. By first order change is meant an improvement in what the organization does or how it does it. By second or third order change is meant a system-wide change in the nature of the core assumptions and ways of thinking and acting. The choice of whether to follow a first or second order change process may be as much determined by organizational politics as by the issues under consideration. How the key organizational actors interpret the forces for change and how they form their subsequent judgement as to what choices they have are important political dynamics.

Defining the desired future

Once a sense of the need for change has been established, the most useful focus for attention is to define a desired future state. This process is essentially that of articulating what the organization, unit or subunit would look like after change has taken place. This process is critical as it helps provide focus and energy because it describes the desires for the future in a positive light. On the other hand, an initial focus on the problematic or imperfect present may over-emphasize negative experiences and generate pessimism. Working at building consensus on a desired future is an important way of harnessing the political elements of the system.

Assessing the present in terms of the future to determine the work to be done

When the desired future state is articulated, you then attend to the present reality and ask, 'What is it in the present which needs changing in order to move to the desired future state?' Because the present is being assessed in the light of the desired future, you are assessing what needs changing and what does not. You may judge that, for the change to effectively take place, a change in current structures, attitudes, roles, policies or activities may be needed. As any change problem is a cluster of possible changes, you may need to group particular problems under common headings, i.e. HRM policies and practices, service delivery, information management, reward systems, organizational structure and design and so on. Then you describe the problem more specifically, and ask, 'Which of these requires priority attention? If A is changed, will a solution to B fall easily into place? What needs to be done first?' This step is about taking a clear comprehensive accurate view of the current state of the organization, involving an organizational inquiry which names:

- the priorities within the constellation of change problems
- the relevant subsystems where change is required
- an assessment of the readiness and capability for the contemplated change.

Another element in describing the present is to describe the relevant parts of the organization that will be involved in the change. This description points to the critical people needed for the change to take place. This is an explicit consideration of the political system and where you draw on your skills as a 'political entrepreneur'. Examples of who needs to be involved might include specific managers, informal leaders, IT specialists and so on. Their readiness and capability for change must be assessed. *Readiness* points to the motivation and willingness to change, while *capability* refers to whether they are able, psychologically and otherwise, to change.

Implementing the change and managing the transition

This step is what is generally perceived as being the actual change process, though as we have seen, preparation for change is equally essential. The critical task is to move from the present to the future and manage the intervening period of transition. This transition state between the present and the future is typically a difficult time because the past is found to be defective and no longer tenable and the new state has not yet come into being. So, in essence, the transition state is somewhat particular, as the old has gone and the new has not yet been realized, and so needs to be seen and managed as such.

There are two aspects to managing this transition state. One is having a strategic and operational plan which simply defines the goals, activities, structures, projects and experiments that will help achieve the desired state. As no amount of change can take place without commitment, the second aspect is a commitment plan. The commitment plan focuses on who in the organization must be committed to the change if it is to take place. There may be particular individuals whose support is a prerequisite for the change and a critical mass whose commitment is necessary to provide the energy and support for the change to occur. The political dynamics of building commitment involves finding areas of agreement and compromise among conflicting views and negotiating cooperation (Fisher and Ury, 1986; Ury, 1991).

It is at the management of transition stage that you are likely to make most use of a group which works with you as the core project team. While you are the one doing the dissertation, this group is an organizational project group which contains both technical competence and hierarchical status to manage the project. Hence, there is a need for you to be able to build and maintain the team (Wheelan, 1999).

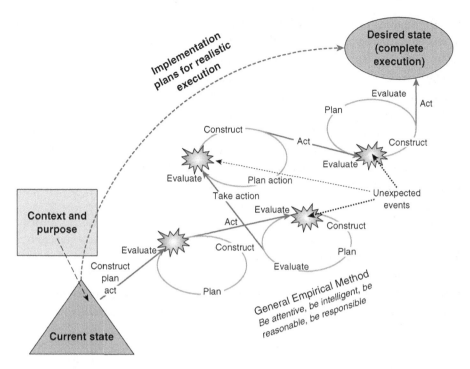

Figure 5.2 Planned change through action research (from the work of Arthur Freedman, with gratitude)

Planned change through action research

Beckhard's (1997) change process developed out of his action research approach and involves continuous interaction between constructing, planning, action and review in order to move a change through a system. Action research has a large degree of messiness and unpredictability about it, in that it is research in real-life action. As the story unfolds, unforeseen events are likely to occur. Environmental events may create a crisis in the organization, fellow key actors may change, and so on. In your role as the insider action researcher and actor-director you are both creating and acting a script. Figure 5.2 illustrates how the path from the current to the desired state and the complete execution of your project is a series of multiple action research cycles of varying complexity, as we described in Chapter 1 through the image of the clock. You are constructing, planning, taking and evaluating action on different issues and on several fronts all along this path. At the same time you are applying the general empirical method of being attentive, intelligent, responsible to the enactment of these cycles.

A good deal of reflection is reflection *on* action, that is, it is a retrospective look at what has happened. As Darling and Parry (2000) illustrate in their application of 'after action review' (AAR) to the US American military, post-mortems can move from being a review of the past to a living practice that anticipates issues and generates emergent learning *in* action.

The critical dimension to action research is how review is undertaken and managed. Review is essentially reflection on experience, and in any such reflection the critical questions are asked not to evoke guilt or blame but to generate learning as to what is taking place and what needs to be adjusted. If review is undertaken in this spirit then the likelihood of individual or team defensiveness can be lessened and learning can take place. As we explored in Chapter 2, learning comes out of reviewing the process and examining emergent questions about content, process and premise.

The action research contribution to the planned change process is that it has collaborative inquiry and learning built in. As Figure 5.2 illustrates, engagement in the cycles of constructing, planning action, taking action and evaluating action provides the structure for shared inquiry into the planned and unanticipated events that occur through the implementation so as to create the conditions and opportunity for learning and for the change to be sustainable.

Deane (2004) was the chief executive of an *ad hoc* government agency, 'Oldorg', whose role was to set, monitor and certify standards for vocational education and training programmes provided within the public further education sector. Under new government legislation, Oldorg was replaced by a new agency, 'Neworg', established in 2001, which subsumed the existing functions of Oldorg and considerably extended its remit. Deane undertook an action learning project to explore how learning could support change in an organization. It involved designing and implementing a learning programme for staff of Oldorg at a time of great uncertainty for the organization, and in a context of vigorous internal and external debate surrounding the introduction of the new legislation.

Deane reports how she adopted Beckhard and Harris's (1987) framework for implementing in Oldorg. A vision of the future had been formulated by examining the given and the possible dimensions of the change arising from the legislation. There had also been an analysis of the present state and an identification of key change issues, focusing in particular on the cultural aspects. What was needed was to complete the process by managing the transition to the changed state. This called for a high level of openness to change across the whole organization. In effect, it called for a dynamic and continuous learning process at both individual and organizational levels to cope with the rate of change in the external environment. While in the past Oldorg had worked through change by learning in a largely intuitive and informal way, Deane now proposed to adopt a more systematic approach. This would closely link the themes of learning and change, first to effect change through learning and second to discover how change could act as a catalyst for learning.

Deane recognized that it was her responsibility to prepare the organization for the future, to predict and plan for change, to influence the change process, and to protect and preserve what was important to the organization in the new situation. She was aware that this presented a great challenge to her leadership: a great deal depended on managing the process effectively, both within and outside the organization.

Having identified the major reasons *why* change was necessary, she moved towards answering the second question: *what* needs to change? She saw how the new organization would have to develop new relationships with stakeholders, serve new customers, adopt new processes and new technologies, and provide new products and services to a bigger market. She reports that she found particularly helpful the notion of a 'constellation' of change issues with a complex set of interrelationships between them. Allied to the notion that it is almost impossible to change only one thing, this concept helped her to frame a change agenda for Oldorg. Focusing mainly on the cultural aspects of change, for which it was suggested by the earlier analysis that the organization's readiness was low, the action learning project then sought an answer to the final question: *how* can we change?

Drawing on her reading on organizational learning and the learning organization, a learning programme was designed and implemented. Simply stated, the purpose of the programme was to ensure that all staff members were given the opportunity to participate in focused learning activities to help them prepare for the changes ahead. The programme comprised a number of learning modes, targeted at producing specific learning outcomes for individuals and for the organization as a whole. Political, technical and culture change issues were explored by means of a learning audit, project teams, action learning groups, staff seminars, a future focus group and training. In terms of culture, the openness of communication and information sharing enabled a high level of ownership of the change. In technical terms, participation in training and new skill development led to new technical systems to prepare for the post-legislation organization. In political terms, the future focus group's submission was accepted and contributed to the development of the process.

Deane reflects on the whole action learning process in first, second and third person terms. For herself, applying the principles of general management to herself and the public sector and being challenged by the action learning setting were the important learning aspects. In second person terms, the organization learned to change and was well adapted to meet the challenges of the new organization. In third person terms, there was a major impact for the clients of the new organization in that they would be able to gain credits for action research undertaken in a work context or elsewhere. The new organization would have a flexible assessment approach to meet the needs of learners.

You may want to read Deane (2004) yourself to gain a fuller picture of her project.

- How does Deane follow the Beckhard and Harris (1987) framework for planned change?
- Are there insights I am receiving from this case that I may bring to my own project?

Learning by design

An alternative approach to the Beckhard change process framework and those developed from it is the 'learning by design' approach (Shani and Docherty, 2003). Learning by design is an approach which builds alternative design configurations through implementing specific learning mechanisms, using action research and other collaborative approaches. Its basic premise is that organizations that prioritize the development and full utilization of their personnel and aim to achieve optimal economic results must explore alternative designs to the bureaucratic organization. These alternatives come about through sustainable learning. Shani and Docherty present six extensive case examples of this approach across different industries, and from them outline a generic eight phase process.

- *Phase 1*: project initiation, that is initial definition of purpose and scope, initial system scanning, securing managerial commitment and role, aligning intervention with business strategy, establishing criteria and measurement of success, reviewing alternative mechanisms to lead effort.
- *Phase 2*: formation of mechanisms to lead the learning by design efforts.
- *Phase 3*: developing a shared vision.
- *Phase 4*: systematic inquiry, analysis and reflection on currently used learning mechanisms.
- *Phase 5*: identification and exploration (of fit) of alternative/additional learning mechanisms.
- *Phase 6*: developing the design of a 'blueprint' for action.
- *Phase 7*: implementation of changes, that is improvement processes for existing learning mechanisms and implementation of new learning mechanisms.
- *Phase 8*: reflection on the 'learning by design' planned change process.

In our view, the learning by design approach can be integrated with insider action research through the process champions or leaders attending to the dynamics in which they engage and how they draw on their preunderstanding, reflect on role duality and manage organizational politics in their efforts to develop learning mechanisms in their own organizational systems. The approach creates capability for human, social, economic and ecological sustainability (Docherty et al., 2008; 2009).

Roth et al. (2007) present their account of insider action research in terms of building learning capabilities and they construct a model of new organizational capabilities in the biopharma firm AstraZeneca, drawing on business strategy, insider action research and learning mechanisms. They note that when experimenting with different ways of creating learning between project teams, they created an arena for reflection and dialogue about operational practices and team behaviours. This intervention became more focused on creating learning within and between teams in a systematic manner. The concept of learning was developed across the organization, supported by a knowledge facilitation method and a team of facilitators. This structure became the preferred way of approaching team learning and became an organizational capability.

You may want to read Roth et al. (2007) yourself to gain a fuller picture of their project.

Questions for reflection and discussion

- What might the setting be for adoption of a learning by design approach to an insider action research project?
- Do these criteria fit your setting and your initiative?

We cautioned earlier against limiting action research to being a logical and clinical process, where individuals and groups move through the action research steps in a rational, albeit it politically aware, manner. However, such approaches are not exclusive. It is not uncommon for researchers to utilize storytelling, drama or photography as a core process of their data generation (Marshall, 1995).

Evans (1997) studied her own practice as a deputy head of a large secondary school in the UK. Within a hierarchically organized institution, she worked with teachers collaboratively, enabling the creation of a learning community through dialogue in which they took ownership of their own development, established a value position and supported one another. She created case studies out of her experiences in the school, and presented them as stories to the group. For example, she composed one story, titled 'Just Tell Me What to Do', out of her own experience. The other teachers were able to relate to the story as it reflected school culture. This and other stories enabled the teachers to reframe their perspectives and to explore new perspectives together.

Questions for reflection and discussion

- What insights do you have of the possible effect of Evans's approach in her school?
- Are there ways in which it would be useful for you to employ techniques such as storytelling, drama and art to give voice to participants in your project?

Data generation as intervention

In action research, data come through engagement with others in the action research cycles. Therefore, it is important to know that acts which are intended to collect data are themselves interventions. So asking an individual a question or observing him at work is not simply collecting data but is also *generating* learning data for both you the researcher and the individual concerned. In short, you are not neutral. Every action, even the very intention and presence of research, is an intervention and has political implications across the system. Accordingly, it is more appropriate to speak of data *generation* than data gathering.

For you, as the insider action researcher, data generation comes through active involvement in the day-to-day organizational processes relating to the action research project. As the researcher in your own organization, you are an inconspicuous observer, as your presence is taken for granted. Your observations are made as a member of the organization in the day-to-day interactions with colleagues and others. Data are being generated not only through participation in and observation of teams at work, problems being solved, decisions being made and so on, but also through the interventions which are made to advance the project. Some of these observations and interventions are made in formal settings such as meetings and interviews; many are made in informal settings over coffee, lunch and other recreational occasions.

You will need to document your reflections for all these occasions. Apart from your journaling activities which typically take place in private, there are the situations at meetings where you want to take notes or jot down reflections. This, of course, is a sensitive process as appearing to take notes may create suspicion. A useful rule of thumb is to adopt what others are doing. If at a meeting most people take notes, then it is okay for you to take notes. If no one is taking notes, then you don't take notes. In that case, you try to jot down your reflections afterwards, as soon as possible, while events are fresh in your memory.

When you observe the dynamics of groups at work – for example, communication patterns, leadership behaviour, use of power, group roles, norms, elements of culture, problem solving, decision making, relations with other groups – you are provided with the basis for inquiry into the underlying assumptions and their effects on the work and life of these groups (Schein, 1999; 2009). As you are dealing with directly observable phenomena in the organizations with which you are working, the critical issue for you is how to inquire into what you are observing and, at the same time, to be helpful to the system. For example, at a team meeting, you may notice all sorts of behaviours which you suspect affect how the team goes about its work: people not listening to each other, wandering off the agenda and so on. If you make an intervention into these areas you are aiming to focus on what is useful for the advancement of the action research project, rather than what you have observed. Without this discipline you may reflect on what you have observed, but the observation may not be owned by participants in the system, because it doesn't meet their needs as experienced or it appears to be showing how clever you are in observing these things. For you, observation and inquiry into how the systemic relationship between the individual, the team, the interdepartmental group and the organization operates is critical to the complex nature of organizational problem solving and issue resolution (Coghlan and Rashford, 2006).

You may consider using some form of survey instrument (Nadler, 1977). An action research approach suggests that data gathering tools need to be designed to fit both the organizational setting and the purpose of the research. While surveying employees by questionnaire as to their views on some aspects of their work or the organization tends to be seen merely as a method of collecting information, it is more important to see how it is an intervention. The reception of a questionnaire by employees may generate questioning, suspicion, anxiety, enthusiasm – all of which are real data for you. If you ignore this you may be missing a key element of how the organizational issue exist and do not get resolved – and indeed what issues lie ahead in the research process.

In a similar vein, interviewing in action research is not simply a tool for collecting data. As we have pointed out, asking someone a question or a series of questions is a data generating intervention. Interviewing in action research tends to be open-ended and unstructured, focusing on what the interviewee has to say rather than confirming any hypothesis the action researcher might have. In Chapter 2 we presented a typology of interview techniques: pure inquiry, diagnostic inquiry and confrontive inquiry (Schein, 1999; 2009). As we emphasized in that chapter, combining inquiry with advocacy is a critical skill for you as an insider action researcher.

While a distinction is made between the study of formal documentation (what Gummesson calls 'desk research') and interviews, Gummesson (2000) makes the point that these are artificial distinctions as the researcher is faced with a continuous flow of data. Secondary data are numeric and textual data that were developed for some purpose other than helping to solve the action research question in hand. You need to evaluate these data on the basis of their relevance to the research question, their availability and their accuracy. In order to have confidence in the worth, validity and reliability of the data, you need to consider the following questions for each archival source:

1 Who collected the data?
2 When were they collected?
3 What was collected?
4 Why were they collected?

Studying relevant documentation can be an important part of organizational research. Access to documentation is integrally linked to the level of access to engage in research. Reports, memos, minutes of meetings and so on may be highly confidential and access to them may depend on the degree to which an organization's management is willing to grant access to the inside researcher. Other documentation may be held in archives in the organization or in libraries. Hill (1993) provides both a general introduction and practical guide to using archives.

The role of technology

Technology has transformed the world of communication. Within organizations, communication through e-mail, websites and intranets have become the norm. Virtual team meetings transcend boundaries of time and space. Information technology (IT) is increasingly becoming the standard method for communication between people and organizations. Technology is also shaping the development of action research processes (Koch, 2007).

The roles that information technology can play in your action research process are manifold. You can use technology to gather information, process it and present conclusions. You can use technology to communicate with your co-researchers, colleagues and clients on a one-to-one basis, to hold virtual group meetings and to communicate with large numbers of people. In regard to this latter use of information technology, an important challenge for you, as the action researcher, is to attend to the quality of participation that occurs through the technology. As Schein (2003) points out, the absence of 'functional familiarity', that is the experience we build up of working with individuals in a face-to-face manner where we know how to read their responses, their body language and general way of interacting, can become a severe limitation on collaborative processes.

How do you know when to stop?

Action research projects which act as dissertations typically have an inbuilt time schedule. Especially within single year or two year master's programmes, you are expected to do your action research project within a designated period in order that you may meet the requirements of the programme in which you are enrolled. Accordingly, you may take your submission deadline, the amount of time you are going to give yourself for writing the dissertation, and work back to where your organizational story will end. In many respects, the decision you make as to when your story will end is arbitrary. At the same time it is important to set a date, after which, whatever takes place, however exciting and relevant, will not be included in your story.

When completion deadlines have more flexibility, your decision to stop is still arbitrary and may depend on your judgement as to the extent that your project has yielded sufficient learning.

Conclusions

In this chapter we have explored how you might go about designing and implementing your action research project. We have shown that, after determining the need for it, it is useful to work at articulating a desired future before getting into details of what to do and how to build commitment to the action. Accordingly, you need to keep in mind that everything you do is an intervention and that you need to be sensitive to the impact that asking questions and observing have on participants. You need to manage the politics of the situation at all times.

Exercise 5.1   The process of implementation

This exercise is for group discussion.

Step 1
Determining the need for change:

- What are the external forces driving change?
- What are the internal forces driving change?
- How powerful are these forces?
- What choices do we have?

Step 2
If things keep going the way they are without significant intervention:

- What will be the predicted outcome?
- What is our alternative desired outcome?

Step 3
What is it in the present that we need to change in order to get to our desired future: what is done, how work is done, structures, attitudes, culture and so on?

Step 4

- What are the main avenues which will get us from here to there?
- What are the particular projects within those avenues (long, medium, short term)?

(Continued)

(Continued)

- How do we involve the organization in this project?
- Where do we begin?
- What actions do we take to achieve maximum effect? medium effect? minimum effect?
- How will we manage the transition?
- How do we build commitment? Who is and who is not ready and capable for change?
- How will we manage resistance?
- Who will let it happen, help it happen, make it happen?
- Do we need additional help – consultants, facilitators and so on?

Step 5

- What review procedures do we need to establish?
- How do we articulate and share what we are learning?

Exercise 5. 2   Reflection for praxis

This exercise is the work of Bob Dick. Copyright ©Bob Dick 1997, 2008. We are grateful to Bob for permission to reproduce it.

Reflection before the action

The '(a)' questions lead to practice. The '(b)' questions lead to theory.

1(a)   What do I think are the salient features of the situation I face?
1(b)   Why do I think that these are the salient features? What evidence do I have for this insight?
2(a)   If I am correct, what outcomes do I believe are desirable?
2(b)   Why do I think that these outcomes are desirable in this situation?
3(a)   If I am correct about the situation and the desirability of the outcomes, what actions do I think will give me the outcomes?
3(b)   Why do I think that these actions will deliver these outcomes in this situation?

Reflection after the action

0(a)   Did I get the outcomes that I wanted? Or more realistically, what were the outcomes that I got, and how well do these accord with those I sought?
0(b)   To the extent that I got them, do I still want them? Why or why not?
0(c)   To the extent that I didn't get them, why not?

These questions return in more detail to the earlier planning questions:

1(a) In what ways, if any, was I mistaken about the situation?

1(b) If I was mistaken, which of my assumptions about the situation misled me?

1(c) What have I learned? What different conclusions will I reach about similar situations in the future?

2(a) In what ways, if any, was I mistaken about the outcomes or their desirability?

2(b) If I was mistaken, which of my reasons for favouring these outcomes misled me?

2(c) What have I learned? What outcomes will I try to pursue when next I'm in such a situation?

Notice that 3(a) takes a different tack:

3(a) Did I succeed in carrying out the planned actions? If not, what prevented or discouraged me? What have I learned about myself, my skills, my attitudes and so on?

3(b) If I did carry out my actions, in what ways (if any) was I mistaken about the effect that they would have? Which of my assumptions about the actions misled me?

3(c) What have I learned? What actions will I try next time I am pursuing similar outcomes in a similar situation?

SIX

Interlevel Dynamics in Insider Action Research

The questions underpinning this chapter are both theoretical and practical. How do you understand the complex interaction of individuals and teams in organizations and in your action research project? How do you work with individuals, with a team and across teams in your project?

Levels of analysis are commonly used as frameworks for researching, understanding and intervening in organizational systems (Harrison, 2005). Levels of analysis typically refer to the identification of issues at units of complexity, such as the individual, the group, the intergroup and the organization. They are important dimensions in action research as traditionally they are seen as targets for research, i.e. action research focusing on the workings of a team. This chapter extends the traditional focus of levels of analysis as the target for research to the notion of levels of aggregation or interlevel dynamics and how interlevel dynamics are important in doing action research in your own organization (Coghlan, 2002). Interlevel dynamics illuminate first, second and third person practice.

Levels of complexity – individual, group, intergroup, organizational – are frequently used as frameworks for understanding organizational processes. Several essential points need to be made about the concept and usage of the term 'levels'. First, the notion of levels must be distinguished from that of echelon (Rousseau, 1985). Echelon refers to position on a chain of command in an organization, such as worker, supervisor, manager, group manager and chief executive. The less common use of organizational levels as a construct in organizational behaviour, however, describes levels of complexity.

Coghlan and Rashford (2006) present levels in terms of how people participate in organizations and link them to provide a useful tool for the manager, consultant and teacher of organizational behaviour. They distinguish four levels of behaviour in organizations: the individual, the face-to-face team, the inter-departmental group and the organizational. The first level is the *matching* relationship that the individual has with the organization and the organization with the individual. For the individual, this involves a utilization of membership and participation in the organization in order to meet personal life goals, while for management the core issue is to get a person committed to the goals, values and culture of the organization so as to perform more effectively. The more complex approach to participation exists in establishing *effective working relationships in a face-to-face team*. An even more complex involvement exists in terms of the interdepartmental group type of interface where teams must be *coordinated* in order to achieve complex tasks and maintain a balance of power among competing political interest groups. Finally, the most complex, from the point of view of the individual, is the relationship of the total organization to its external environment in which other organizations are individual competitors, competing for scarce resources to produce similar products or services. The key task for any organization is its ability to *adapt* to environmental forces driving for change.

Viewing organizations through levels of analysis is only one part of the picture. The other part refers to how levels are related to each other. In an organization there is an essential interlevel element in that each level (the individual, the group/team, the interdepartmental group, the organization) has a dynamic relationship with each of the others (Coghlan and Rashford, 2006). This relationship is grounded in systems dynamics, whereby the relationship each level has with each of the others is systemic, with feedback loops forming a complex pattern of relationships (McCaughan and Palmer, 1994; Haslebo and Nielsen, 2000). For instance, dysfunctions at any level can lead to dysfunctions at any of the others. An individual's level of disaffection may be expressed in dysfunctional behaviour in the team and affect a team's ability to function effectively, which in turn reinforces the individual's disaffection. If a team is not functioning effectively, it can limit the interdepartmental group's effectiveness, which may depend on the quality and timeliness of information, resources and partially completed work from that team. If the interdepartmental group's multiple activities are not coordinated, the organization's ability to compete effectively may be affected. In systemic terms, each level affects each of the others. Viewing organizational levels as simply 'levels of analysis', without taking interlevel dynamics into consideration, misses the point about the systemic relationship the individual has with the team, the team with the individual, the team with other teams, the organization with its environment and each with each other.

When you are working at each or any of the above levels, you will typically find that you are challenged to include the other levels in your diagnosis and intervention. For instance, you may be working with a team on an aspect of your action research project. In the process of this work, it emerges that some of the individual members are experiencing dissatisfaction with their relationship with the organization and don't provide an optimal contribution to the team's endeavours. Or it may be that the flow of information from other teams is having a negative effect on the work of the team with which you are working. In these instances, you are challenged to move beyond the team level intervention in which you are engaged to consider dynamics at the other levels that are having an impact on the team.

Organizational levels are important dynamics in organizational politics. Organizational political behaviour may be individual, team and interdepartmental group. Individuals may engage in covert political behaviour in order to advance their own standing in the organization. Teams may engage in overt or covert political behaviour to gain advantage over other teams in order to obtain more resources.

Interlevel dynamics of change

Organization change typically means that individual employees have to change. Individuals may be required to change what they do or how they do it. It may be required of them to change their attitudes towards their work or some particular aspect of it. A consequence of this may be that an individual's relationship to the changed or changing organization may be altered, either positively or negatively. An organizational change, if not well managed, may result in individuals feeling alienated. A change agenda in an organization typically affects the work of teams. Teams may set new goals and targets; they may have to work differently. As resources are reallocated and technology and advanced information systems alter access to the flow of information, teams are required to communicate more effectively across functions and departments. So we can see that organization change comprises individual change, team change and interdepartmental group change.

What happens at one level has an impact on each of the others. So, for example, a recession or slowdown in the global economy leads to a reformulation of strategy, a realignment of operations across the interdepartmental group, a change in the work of teams, and change for individuals. Some individuals may be laid off and so their relationship to the organization ceases, while others may benefit through reskilling and professional development. If a team is working well, that contributes positively to the motivation and participation of the individual members; the reverse is also true. The work of one team affects the work of others. So each level is systemically linked to each of the others, and events at one level are both cause and effect of events at other levels.

There are critical interlevel dynamics in the process of organizational change that we described in the previous chapter. The change process has to begin somewhere and typically it begins with an individual, though not exclusively at the top. For the change agenda to progress, that individual has to take it to a team and that group needs to adopt the need for change. When a management team adopts the need for change and begins to act, it has then to win over other teams in the system. Each of these movements – from individual to team to other teams – is an iterative process. In other words, when the team adopts an individual's position, that adoption reinforces the individual; when other teams adopt a particular team's position, that reinforces that team; and, of course, when customers adopt a new product or service, that reinforces the organization. The process of defining the future involves interlevel dynamics. If the vision comes from the chief executive, then there are interlevel dynamics from that individual to the senior management group and then to the other teams and on to the organization. So the iteration of issue presentation, reaction and response ebbs and flows from individual to individual, team to team and so on. The process of designing the vision of the future involves interlevel dynamics. If the vision does not come from the chief executive, then there are interlevel dynamics from that individual to the senior management group and then to other teams and on to the organization. So the iteration of issue presentation, reaction and response ebbs and flows from individual to individual, team to team and so on. Interlevel dynamics are pivotal to the processes of the transition state as individuals and teams address the implications and implementation of the change agenda. As the change agenda affects the work of individuals in what they do and how they do it, individual commitment is essential. As the change agenda affects the work of the permanent teams and typically requires the creation of new teams and work in temporary committees or project groups, team dynamics are critical to the change process. In a similar vein, the change agenda involves the interface of multiple teams with respect to information sharing, problem identification and resolution, resource allocation and collective bargaining; thus interteam dynamics can enable or hinder the successful management of the change process.

Interlevel dynamics of strategy

In the strategy process, interlevel dynamics abound and are important processes with which you may work as an insider action researcher. The complex process of generating strategic thinking and acting evolves through five strategic foci (Coghlan and Rashford, 2006):

- framing the corporate picture
- naming the corporate words
- doing corporate analysis
- choosing and implementing corporate actions
- evaluating corporate outcomes.

The first focus, *framing the corporate picture*, deals with the key individuals and the history of the organization as it comes to affect strategy formulation and implementation. The contents of this focus are the core mission and its statement of the organization as well as the characteristics of the key players. If the corporate picture comes from the senior manager, then he/she works to get the senior management group to accept it and then the interdepartmental group and ultimately the organization – stakeholders, shareholders and customers. A new corporate picture forces the senior management group to set new priorities and allocate work to the divisions and departments in a different way. Interdepartmental issues centre on the new allocation of resources and information flows required in carrying out the new corporate picture.

The second focus, *naming the corporate words*, deals with those operations or functions that had become the lead operations or functions over time. These are most often referred to as driving forces and become the articulation or lived-out mission statement of the organization. There are interlevel dynamics in choosing the corporate words as functional areas compete for priority and status. For example, those who are not in engineering or marketing, where one or other is defined as the prime function, may feel undervalued and peripheral in terms of their contribution to the organization.

The third focus, *doing corporate analysis*, deals with the obtaining of critical information and the preparation of these data into a comprehensive scenario or modelling of alternatives for the organization. The analytic process acknowledges the appropriate connectedness of this focus to the previous focus of mission in framing the corporate picture and to the subsequent focus of choosing and implementing corporate actions which will of necessity follow. There are cultural issues in how an organization engages in analysis. Each functional area has to examine its own analysis and process its own functionality as well as how the interdepartmental group's function affects others and its output. Each functional area is a subculture, which exists in an organization where functional groups have developed their own traditions, language and basic assumptions, which provide a self-contained subsystem. So each functional level is offering its view of the whole from the perspective of its own subculture in the interdepartmental group activity. How these analyses or 'views' interrelate or are accepted is critical as most often, in integrating multiple functions, the cross-links between functions are the sources of trouble.

The fourth focus, *choosing and implementing corporate actions*, deals with the selection and structured implementation of a strategic plan of action with its concurrent implementation plans. In choosing and implementing corporate actions, the role of the final decision maker resides with the CEO. CEOs interface with a senior management group. When this interface is prolonged and the differences between alternatives are ambiguous at best, the CEO can get trapped into deciding along the lines of his or her own past career experience.

The fifth focus, *evaluating corporate outcomes*, deals with the acceptance of the choice of criteria in the appropriate review and evaluation of the resulting state of the organization. The process aspect of evaluating corporate outcomes looks at the appropriate and fair evaluation of outcomes. The evaluation of the corporate outcomes most often resides with the senior management group and the CEO.

Evaluation of both outside and inside reviews provides the grist of the interlevel interaction between the senior management group and the CEO. This discussion focuses on the nuances of how the data are formed into information. The senior management group views the data from its interdepartmental perspective while the CEO looks at the data most often from the total organization's point of view or even an outside point of view.

Each of the five strategic foci entails interlevel dynamics of change as an organization seeks to frame its corporate picture, name its corporate words, do its analysis, choose and implement corporate actions, and evaluate outcomes. But experience tells us that these levels are not only discrete and separate but also interdependent and interrelated.

Levels of analysis in action research

Action research at the individual level

In Chapter 2 we explored how you can engage in first person practice by attending to your own learning in action. First person research in your own organization, that is how the research is for you, is linked to your own sense of *matching* to your organization. As we have seen, how the research contributes to your own development, your role and your future in the organization is a significant aspect of undertaking research in your own organization. Accordingly, your own self-awareness, your closeness to the issues, how you frame the issues and so on are critical first person processes of which you need to be aware and to work on consciously as part of the action research project. Individual learning in action typically involves being able to reflect on experience, understand it and enact chosen alternative behaviours and to learn to critique your assumptions in a manner which exposes your private inferences to public testing. The degree to which the project enhances your career or to which it decreases your motivation to remain in the organization is important.

Action research at the group or team level

Second person research is characterized when researchers engage with others in conversation and action. This may be actualized in one-to-one situations where you engage in action and reflection with a single individual. More frequently, second person practice is enacted in groups, whether in the formal organizational hierarchical teams or the temporary committees or task forces that are built around the project. They may be collegial groups which you have set up to explore the task of your action research project, such as a cooperative inquiry or an action learning group.

The experience of groups and teams in engaging in the action research steps is paramount. As they engage in the activities of diagnosing, planning and taking action they may experience some success in some of their activities and not in others. They may experience internal conflict and destructive political behaviour by some members. They may struggle to reach agreement on strategies and action and so on. What is important is that groups and teams learn to reflect on their experience in terms of how they function as groups and teams. This involves attending to task issues of how they do the task and relational issues of how they manage communication among themselves, solve problems, make decisions, manage conflict and so on (Schein, 1999). Exploring these issues means being able to go beyond personal blame and draw on useful constructs on effective group and team development to take remedial action where necessary and develop effective team processes (Wheelan, 1999). These activities involve content, process and premise factors as the issues on which they are working are studied and the ways in which the teams work are reviewed and the underlying assumptions are uncovered and examined. Bartunek (2003) provides an illustration of the development of change agent teams over time in terms of identity, actions and stakeholder relationships. She narrates cognitive and affective links between (1) how identity evolved over time in the face of member turnover, explicit reminders of identity and face-to-face contact; (2) how actions intended were taken and how emotionally engaging the actions were; and (3) how efforts to make these stakeholder relationships positive were taken.

Action research at the intergroup level

Yet the research and change process cannot be restricted to learning and change by individuals and teams alone. A further application of second person research is how the learning and change which take place in individuals and teams need to be generalized across the interdepartmental group, whereby other teams and units engage in dialogue and negotiation. A critical focus for attention in this regard is the impact that cultural perspectives from different functions have on the change process (Schein, 2004). Intergroup dialogue needs to take account of how functional areas in organizations hold different assumptions from and about one another.

Groups and teams do not work in isolation. They are typically members of wider systems and such membership involves intergroup dynamics, such as being interdependent in a work flow or information process. The consequences of intergroup interdependence are intergroup dynamics, such as the sharing of information, negotiation over resources, intergroup prejudice and intergroup conflict. You can expect to have intergroup dynamics in any project involving a complex system. Indeed, working at the intergroup level is an extension of second person research. The interdepartmental group experiences the differences between groups, as different groups are separated from one another by what they do, by location and by their interests. Accordingly, any action research work which involves members of separate departments working together must take account of how each department has its own concerns, its own view of the world, its own political interests in the work of the project, and even its own terminology and language. We would argue that interdepartmental group work is essentially intercultural (Schein, 2004).

A further instance of action research at the intergroup level is found in the burgeoning use of large group interventions in organizational change. Large group interventions are described as search conferences, future search, dialogue conferences, open space and real-time strategic change, among others (Gustavsen and Englestad, 1986; Bunker and Alban, 2006; Holman et al., 2007; Martin, 2008), and are gatherings of the members of a system in a large group in order to create common ground for the future development of that system.

Action research at the organizational level

Finally, the action research project does not stop within the organization. Action research at the organizational level means that the project encompasses the organization as an entity in a competitive economic and social environment. Organizations as open systems have a dynamic two way relationship with their external environments (Katz and Kahn, 1978). The second person research process includes how the organization is affecting and being affected by its customers or clients, stakeholders, local community, competitors, wider society and other organizations. The project therefore is inclusive of the organization's relationship with its external stakeholders such as customers, clients and competitors, as well as with internal stakeholders at the individual, group and intergroup levels.

To add further complexity, the project may involve interorganizational work, typically called interorganizational networking. Interorganizational networking is where member organizations deliberately develop voluntary networks to help deal with complex issues and devise collaborative ways of planning and taking action (Coghlan and Coughlan, 2005; Burns, 2007). The learning steps of the action research cycle need to be inclusive of reflection on how different mindsets and political interests in different organizations and groups experience working together, how they process and interpret that experience and how they take action accordingly.

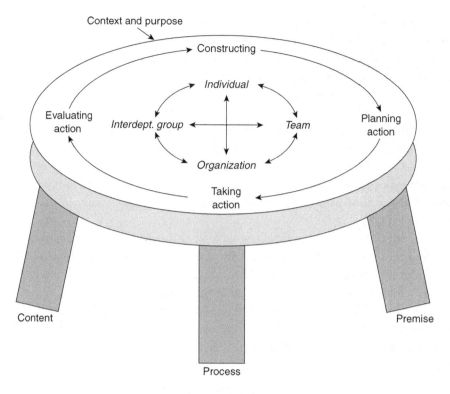

Figure 6.1   The Organizational Dynamics of Action Research

Conclusions

Levels of analysis are commonly used as frameworks for researching, under-standing and intervening in organizational systems. Levels of analysis typically refer to the identification of issues at units of complexity such as the individual, the group, the intergroup and the organization. They are an important dimension of action research as traditionally they are seen as targets for action research. This chapter reviewed the traditional focus of levels of analysis as targets for action research and extended the notion to levels of aggregation or interlevel dynamics.

Many of the case examples cited through the book, while appearing to focus on one unit of analysis, actually contain interlevel dynamics. Krim (1988), while focusing on own learning, was embroiled in complex interlevel dynam-ics in his engagement with key political players and groups in city hall. In OilCo, the CEO formed his managers into an executive council (Kleiner and Roth, 2000). The managers had not worked together as a team before, so it was a new experience for them to do so. They formed subteams to work on partic-ular projects. The executive council disseminated its work to the wider organi-zation through a learning convention, at which people spoke frankly for the first time about their views of the company and dialogue began to take place.

The process continued through the company, whereby the transformation of fundamental ways of thinking and feeling moved from the CEO to the executive council and through the organization and impacted teams and individuals.

Interlevel dynamics are also operative for you as an action researcher in the action research process itself (Figure 6.1). You engage in interlevel dynamics as you encounter the process of the complex systems in which you enact action research projects. Your engagement in first and second person research typically requires enactment of your own learning in action along with work with groups, between groups and with organizations in their environments. The application of interlevel dynamics to first and second person research is an important dimension of both understanding and enacting the experiential cycles of action research.

You may use the construct of the four levels as a diagnostic framework by being aware of the issues occurring at each level and how one level affects another, and being able to work with individuals, teams and interteam groups to evaluate the effect of one level on another. For instance, the process of moving a change through an organization requires a systemic view of the complex interrelationship and interdependence of the individual, the face-to-face team, the interdepartmental group and the organization.

Interlevel dynamics are systemic processes. They provide frames for us to understand how participation in human systems is developed through increasing complexity: individuals, individuals in groups and teams, individuals in groups and teams which are part of an interdepartmental group with other groups and teams, and the interdepartmental group within an organization, which itself is a participant in a sector, a market and the global economy. These frames are not only for the purpose of understanding but are also the basis for action (Coghlan and Rashford, 2006). Hence, your research in action can benefit from an awareness of and skills in working with interlevel dynamics.

Exercise 6.1   Applying interlevel dynamics

Consider a project that you are planning to undertake in your organization.

1 (a) Who are the individuals involved in this project? How do I work with them?
   (b) Who are the teams involved in this project? How do I work with them?
   (c) What are the issues between these teams? How do I work at the interteam level?

2 In the teams in the project, what impact are individuals having on the team in which they are members and vice versa?
3 What are the significant patterns of relationships between individuals, teams, the interdepartmental group and the organization that I need to be sensitive to and work with?

Exercise 6.2   Change issues

	Individual	Team	Interdepart mental group	Organization
Individual				
Team				
Interdepartmental group				
Organization				

1 Name a change issue with regard to an individual.
2 Now work diagonally along the unshaded boxes. How does one level have an impact on the others?
3 Where would you put your energies to advance the progress of the change and heal dysfunctions?

1 Name a change issue which applies to the whole organization.
2 Now work diagonally along the unshaded boxes. How does one level have an impact on the others?
3 Develop strategies to implement change at each level.

Exercise 6.3   The learning window

The learning window is an adaptation of the famous JOHARI window and is created by Lyle Yorks, Teachers College, Columbia University, New York. We are grateful to Lyle for permission to use it.

Learning window

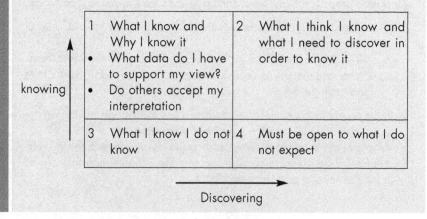

1 What I know and Why I know it • What data do I have to support my view? • Do others accept my interpretation	2 What I think I know and what I need to discover in order to know it
3 What I know I do not know	4 Must be open to what I do not expect

knowing ↑

Discovering →

The learning window can be used in an action research group. It aims at enabling the group to distinguish between what it knows and what it is inferring and thereby acting on the basis that it thinks it knows. Making these distinctions helps keep the group focused on data.

- *Pane 1*: what the group knows has to contain solid data that have been tested and meet with consensual agreement among group members.
- *Pane 2*: what the group thinks it knows catches the inferences and attributions that group members are making and challenges the group to make those inferences explicit, to locate them in directly observable behaviour through the ladder of inference, and to see them as hypotheses to be tested rather than accept them as facts.
- *Pane 3* identifies the gaps in knowledge that the group knows it needs to address and opens up an agenda for further data collection and hypothesis testing in action.
- *Pane 4* is the blind area where the group does not know what it does not know.

Fill in the panes of the blank learning window below with your group. Discuss what goes into each pane and the evidence that you have that locates information in each particular pane.

1 What we know	2 What we think we know
3 What we know we don't know	4 What we don't know that we don't know

SEVEN

Using Frameworks to Study Organizations in Action

We now turn our attention to the process of making sense of organizational dynamics by presenting some features of the use of frameworks in understanding organizational dynamics. The underpinning question is: how do I use frameworks to make sense of what I am seeking to understand? Organizational dynamics and the use of frameworks are subjects in themselves (Harrison, 2005; Bolman and Deal, 2008). There are innumerable frameworks, which can be found in standard textbooks and in the writings of those authors who have created them. For instance, within the field of business strategy, you may be familiar with frameworks which enable you to analyse the competitive nature of an industry or the relative position of a firm within an industry. Within marketing, you may draw on frameworks which help position a product or service. In every field and subject area there are frameworks which enable you to make sense of the current situation and which help you predict outcomes. In relation to doing action research in your own organization, we highlight the systems approach and give particular attention to constructs of organizational learning and change. We are not attempting to provide a list of such frameworks; rather we are aiming at providing an introduction to their use.

Organizational diagnosis

We use the term *diagnosis* to refer to investigations that draw on concepts, models, and methods from the behavioural sciences in order to examine an organization's current state and help clients find ways to solve problems or enhance organizational effectiveness. (Harrison and Shirom, 1999: 7)

Underlying the principle of organizational diagnosis is a notion of organizational health which organizational clinicians are using to compare with the present situation (Schein, 1997). Accordingly, frameworks which postulate key organizational variables and relationships are important diagnostic tools. At the same time, we remind you of our important point in Chapter 1, namely that we do not see a 'diagnosis' as an activity aimed at creating an objectivist truth. Rather, we understand the use of frameworks as providing a construction for conversation and mechanisms for collaborative sense making and joint action planning and action.

Organizational frameworks are presentations of organizations which help categorize data, enhance understanding, interpret data and provide a common shorthand language (Burke, 2008). They typically describe relationships between organizational dynamics, such as purpose, strategy, structure, control systems, information systems, rewards systems and culture, and help organize data into useful categories and point to what areas need attention.

Some guidelines are useful for selecting and using frameworks. Weisbord (1988) advises that frameworks should have four features: that they be simple, fit members' values and highlights things they consider important, validate members' experience by putting recognizable things in a new light, and suggest practical steps. Burke (2008) provides three guidelines for selecting a framework. The first is that you should adopt a framework you understand and with which you feel comfortable. The second is that the framework selected should fit the organization as closely as possible, be comprehensive enough to cover as many aspects of the organization as appropriate, and be clear enough for members of the organization to grasp. The third is that the framework should be sufficiently comprehensive to enable data gathering and interpretation without omitting key pieces of information. In a word of caution, Burke points out that you may become trapped by your frameworks, so that your way of seeing becomes a way of not seeing. So as an action researcher, you need to critique the frameworks you use.

Systems thinking and practice

A significant contribution to situation analysis is systems thinking and practice. Systems thinking refers to seeing organizations as a whole, made up of interrelated and interdependent parts. The human body is a good example of a system, whereby bones, muscles, tissues, and organs perform interdependent and interrelated functions. While we might dissect the body and make an analysis of any particular part, the body's functioning depends on a holistic view of how all the parts work together. Similarly, organizations may be viewed as systems, in which planning, control, structural, technological and behavioural systems are interdependent and interrelated.

Understanding organizations as open systems, that are dependent on the external environment for their survival, has been well established in organizational theory for many years (Katz and Kahn, 1978). What has received less emphasis is the 'recursive' systems model, which represents organizations as patterns of feedback loops and sequences of interaction which link and integrate elements of a system (Senge, 1990; McCaughan and Palmer, 1994; Haslebo and Nielsen, 2000). In systems thinking, linear cause and effect analysis is replaced by viewing patterns of interaction which mutually influence each other.

'Dynamic complexity' refers to situations where a system is complex, not because of a lot of detail but because of multiple causes and effects over time (Senge, 1990). In situations of dynamic complexity, systems thinking provides a perspective of viewing and understanding how a system is held together by patterns of action and reaction, relationships, meanings and hidden rules, and the role of time. In order to inquire into how a system functions, you can engage in systemic questioning (McCaughan and Palmer, 1994):

- *Establishing circuitry*: when A does X, what does B do? What does A do next?
- *Establishing patterns*: what patterns are evident over time?
- *Exploring meaning*: what are the meanings held in the system? What are the common meanings attributed to events and actions?
- *Exploring covert rules*: what unarticulated and hidden rules govern behaviour?
- *Exploring the time dimensions*: how time delays have an impact on the system.

It is not easy to find answers to questions posed by systemic questioning. Formulation of tentative working explanations as to what is happening in the system – the circuitry, patterns, covert rules, meanings and time – may uncover the dynamic complexity of the system, and may involve many iterations of collaborative inquiry before explanations that fit are found.

Campbell (2000), in his book on the socially constructed organization, makes the point that in the early 1980s a paradigm shift took place from what was traditional general systems theory, which is referred to as first order cybernetics, to second order cybernetics, which places the observer within the system he/she is observing. The emphasis shifted towards the constructs that the observer was bringing. Rather than viewing the system as something connected by feedback and difference, there was a shift to understanding the system as generating meaning. Social construction seeks to grasp how people work together to produce their realities. It distinguishes itself from systems thinking by moving to questions about what is happening to questions about how it happens, and from observations of action to explanations of action. It also poses questions about who is included or excluded in conversations, how the conversations are run and what they are about (Bushe and Marshak, 2008).

Systems thinking and action research

As an insider action researcher, you are seeking insight into the systemic patterns of thinking and acting in your system but also into the constructs and meanings you bring to your inquiry. Hence, we offer systems thinking as a way of seeing the system, seeing where you are in it, and engaging in collaborative dialogue about how multiple constructions affect what happens and how what happens affects the constructions and meanings that people ascribe to their intentions and actions. Viewing the organization as a social construction is itself a social construction.

Systems thinking and the action research cycle play complementary roles. In a systems approach to action research, tentative explanations are being formed as the story unfolds. These insights are tentative frames to articulate the elements of the system in order that they may be understood and to consider interventions to change them, where required. A very useful way of formulating systemic explanations is through the use of diagrammatic representation. A diagrammatic representation is a powerful way of expressing insight into how a system operates. When cycles of action and their consequences are drawn in a diagram, the insight into the patterns of the system may be illuminated. Both the act of attempting to represent the system diagrammatically and the diagram itself are essential elements of the learning process. The very act of drawing the system's diagram is a learning process of explanation formulation and testing (Anderson and Johnson, 1997). In traditional research approaches, intuition is frequently placed against reasoning and considered alien from a research process. In Senge's (1990) view, the systems approach holds the key to integrating intuition and reason, because intuition goes beyond linear thinking to recognize patterns, draw analogies and solve problems creatively.

Change and learning

As change and learning are central to action research, it is important for the action researcher to draw on knowledge of how change and learning take place (Burke, 2008; Schein, 1996). How change and learning take place applies not only to individuals, but also to groups, between groups and to organizations, as we saw in the previous chapter. Change theory has evolved from Lewin's (1948/1999) model that the change process has three stages or sets of issues: being motivated to change, changing, and making the change survive and work. Lewin argues that a system must unlearn before it can relearn and that attention to all three stages is equally critical.

Any action researcher in an organization needs to understand how people in organizations can resist change. An important starting place is that resistance is a healthy, self-regulating manifestation which must be respected and taken seriously by the action researcher. Coghlan and Rashford (2006) present two psychological reactions to the initiation of a change. When a change agenda is first presented, people may *deny* its relevance. When denial is no longer sustainable it may be replaced by *dodging*, which is an effort at diverting the change. Denial and dodging are natural reactions to a change agenda, especially when it is unexpected. In Coghlan and Rashford's view, they are a prelude to *doing* and *sustaining*, when the change agenda is accepted and implemented.

There are different levels of change and learning which have a particular relevance to action research. From the work of Bateson (1972) and others who have developed his work, a distinction between change or learning which deals with routine issues, and that which involves a change of thinking or adoption of a different mental model, is typically defined as a distinction between single and double loop learning (Argyris and Schon, 1996) and first, second or third order change (Coghlan and Rashford, 2006). *First order change* occurs when a specific change is identified and implemented within an existing way of thinking. For example, Bartunek et al. (2000) describe management-led action research in a bank on a problem of communication failings with clients. Through the action research process of participative data gathering, data analysis, feedback and action planning, intervention and evaluation, the named problem was addressed and improvements were made. *Second order change* occurs when a first order change is inadequate and the change requires lateral thinking and questioning and altering the core assumptions which underlie the situation. In another example, Bartunek and colleagues describe a manager-led action research project which initially aimed at addressing improving a manufacturing system by increasing volume while maintaining flexibility as well as enabling automated material control and improved planning. As the data were being analysed, it became evident that these changes would involve creating a radically new way the company did business. Accordingly, through the action research cycle, materials personnel, assemblers, testers and supervisors/managers participated in diagnosis, analysis and feedback, resulting in the implementation of a new integrated manufacturing system. Due to the success of this project, a similar methodology was applied to other change projects in the company. It is realized that sometimes concrete problems are symptoms of complex attitudinal and cultural issues which must be addressed and that problem resolution involves organizational transformation. This is called *third order change*, which occurs when the members of an organization learn to question their own assumptions and points of view and develop and implement new ones.

Issues may not be obvious, as we discussed in Chapter 4. First order problems may persist unless there is second order change. A recurrent demand for second order change may point to the need for the development of third order skills. Observation of a group at work may yield questions as to what particular behaviours or patterns of behaviour mean. What is critical is that you as the action researcher inquire into those patterns and facilitate the group in surfacing and examining them, rather than making a private interpretation which is untested but then becomes the basis for action. Taking what is directly observable into the realm of meaning requires skills in inquiry and intervention, as we discussed in Chapter 3.

Conclusions

In this chapter we have outlined how you would approach making sense of complex organizational data. We have presented some major themes with respect to how you might go about choosing a framework on which you would base your understanding of organizational data and on the basis of which you would take action. It might be expected that, in this chapter, we would discuss topics like data analysis, discourse analysis and other important techniques from the field of qualitative approaches to research. While some action researchers do engage in this form of analysis out of having conducted interviews and engaged in participant observation activities as part of their action research project, we have refrained from such a discussion. In our view, we do not wish to distract you, as an insider action researcher, from the primary focus on describing, explaining and engaging in action. In our experience, a lack of familiarity or discomfort with the action research approach as presented in Part I of this book (a lack of familiarity or discomfort may apply to academic supervisors as well as to students) may tempt you to take refuge in these techniques.

Making sense has different applications in different contexts where the action research in your own organization is linked to academic accreditation. In a master's programme, such as an MBA or its equivalent, frameworks such as those discussed in this chapter are used to help you see more clearly what is going on and to design appropriate interventions to deal with the issues identified. In a master's by research and a PhD, you go further. In this context, you not only use the frameworks to help you see what is going on and to plan further action, but also critique and extend theoretical frameworks in order to contribute to theory development.

Exercise 7.1 Diagnosing your organization

Take any relevant organizational diagnostic framework from any textbook, and apply the categories and causal links to your own organization.

1 What picture is emerging?
2 What do you need to do to check the picture you have of the organization?
3 Where do you think you need to intervene?
4 How do you justify that diagnosis and intervention selection?

Exercise 7.2 Using systems thinking

Some useful books which can help you do some systems thinking and mapping are Senge et al. (1994) and Anderson and Johnson (1997).
  On a sheet of paper:

1 Describe the issue/problem as you see it.
2 Tell the story.
3 Draw a map of the story: when A said/did X, what did B do? What did A do next? What was the outcome for C?
4 Connect the process of the story with arrows.
5 Include where you are in the story and what your interests are.
6 Consider any number of explanations of the patterns.
7 Consider any number of interventions which might change the structure of the system, and see how each intervention has different outcomes across the system.

Part 3

Issues and Challenges in Researching Your Own Organization

EIGHT

Researching Your Own Organization

Parts I and II have enabled you to enter into your insider action research project and to advance on your action and your inquiry into that action. We now turn to reflect on the specific topic of this book, namely doing action research in your own organization. The questions underpinning this chapter are: what are the implications of engaging in action research in your own organization? What are the particular dynamics that accompany insider action research that you need to understand and to take into account? The following three chapters explore the issues and challenges that you face as you take on your exciting venture. Smyth and Holian suggest that insider research can be a little like abseiling, and that if you have abseiled 'you would know the feeling when you defy gravity, lean back into the empty space parallel to the ground and step off a cliff face' (2008: 42).

Researching your own organization involves undertaking research in and on your own organization while a 'complete member' (Adler and Adler, 1987). As 'permanent' is a term increasingly less applicable to today's workplace, we are using 'complete member' to refer to being a full member of your organization and wanting to remain a member within a desired career path when the research is completed.

The 'complete member role', as outlined by Adler and Adler (1987), is closest to researchers studying their own organization. Such researchers have an opportunity to acquire 'understanding in use' rather than 'reconstructed understanding'. Riemer (1977) argues that rather than neglecting 'at hand' knowledge or expertise, researchers should turn familiar situations, timely events and/or special expertise into objects of study. This orientation was partly abandoned by ethnographers during the 'classical era', when participant observation replaced the life history and the emphasis shifted toward greater objectivity and detachment. Participation and, for some, the research process involved becoming a temporary member of the organization in order to observe at first hand how life was lived; this was accepted and accorded legitimacy, but subjectivity, involvement and commitment were thrust aside.

What is central to this book is how complete members may undertake action research in and on their own organizations, with completeness being defined as before in terms of wanting to remain a member within a desired career path when the research is completed.

In Chapter 3 we provided a brief introduction to the notion of collaborative management research, which involves shared exploration between practitioners and researchers to improve the performance of a system and to add to the broader body of knowledge in the field of management (Shani et al., 2008). The collaboration of practitioners and researchers is paralleled by collaboration between insiders and outsiders. While it might be generally assumed that the insider is the practitioner and the outsider is the researcher, this is not necessarily the case (Bartunek, 2008). The insider perspective is integral to the process of collaborative management research and, as Coghlan and Shani (2008) explore, the challenge is to create a community of inquiry from the collaboration of practitioners and researchers and insiders and outsiders. We invite you to note here that the issues that we explore in this and the following chapters (Chapters 8, 9 and 10) pertain not only to the context of this book, i.e. you as an individual undertaking insider action research, but also to the insider perspective of collaborative management research.

The focus of the researcher and the system

Doing action research in your own organization is opportunistic, that is, you may be selecting an issue for research which is occurring anyway, irrespective of whether or not your inquiry takes place. We have described these as the *core* action research project and the *thesis* action research project (Zuber-Skerritt and Perry, 2002). Hence, we need to distinguish between the two projects, as respective responsibilities may differ. For instance, we know of a case of insider action research in which the researcher is, in effect, doing action research on a major project for which she is responsible. In this case, her academic supervisor has challenged her to differentiate between the actions of the project and the quality of her inquiry into how that project progresses and what knowledge can be extrapolated. Her research is evaluated on the quality and rigour of her inquiry, rather than on the extent and success of the organizational project that she manages and for which she is accountable to her organizational superiors. In contrast, we know of another case in which the insider researcher is working as an internal facilitator in a change project, but is not responsible for its overall management, a role assigned to a senior project manager.

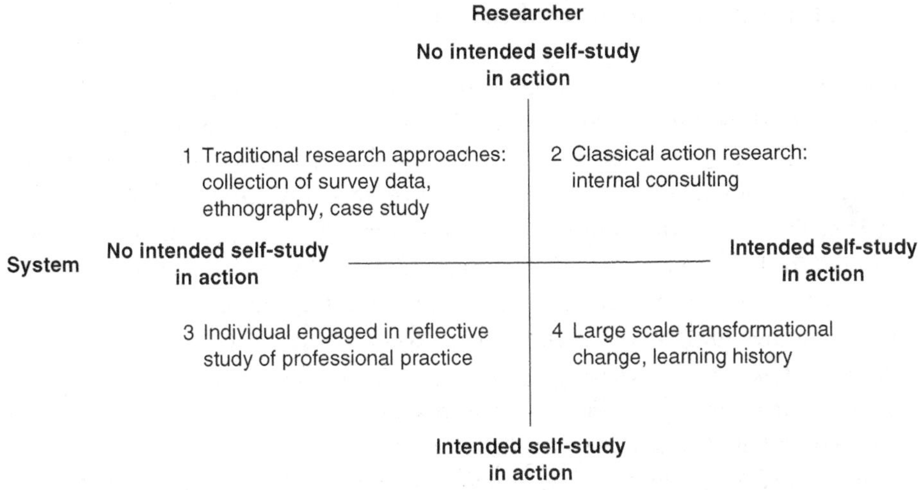

Researcher

No intended self-study
in action

1 Traditional research approaches:
collection of survey data,
ethnography, case study

2 Classical action research:
internal consulting

System · No intended self-study
in action —————————————————— Intended self-study
in action

3 Individual engaged in reflective
study of professional practice

4 Large scale transformational
change, learning history

Intended self-study
in action

Figure 8.1   Focus of researcher and system

Accordingly, we need to differentiate between the researcher and the system
in and on which the action research is taking place, whether that system be
a large organization, a community, a department or a unit. We can reflect on
the intended goals of both the researcher and the system. As we reflect on
cases we know of insider research projects, we notice that the focus of the
researcher and system can vary. For instance, we know of an individual manager whose master's action research project was about the organizational
change he was leading. His second person intervention work to manage the
politics, the power dynamics and the conflicts between key protagonists was
central to both his managerial role in leading change in his organization and
his action research dissertation. His reflection in action was central to his dissertation. At the same time, the members of his organization had little consciousness of the fact that he was doing a dissertation for a postgraduate
degree; in their eyes he was simply doing his job. We know of another case
of an individual, in the same master's programme, who studied how his organization managed information. In this case, the individual's research focused
on what was happening around him and was of great interest to his superiors and other members of the organization, but it did not involve him in any
form of deliberate self-reflection in action.

Given there can be a range of foci on the part of both the researcher and
the system, can these foci be captured in a useful way? We understand that
research can be viewed along a continuum which reflects the intended focus
of the research for both researcher and system (Figure 8.1). We are distinguishing a commitment to intended self-study in action by either or both
researcher and system from no such commitment.

Quadrant 1: traditional research

Quadrant 1 of Figure 8.1 is defined by the absence of intended self-study in action by both researcher and system. There is study but it is not deliberately in action. This is a situation where the researcher is focusing on a perspective or issue or problem within the system as if external to himself/herself, and is not engaging in any deliberate self-reflection in action as part of the research process (Anderson et al., 1994; Flyvbjerg, 2001). At the same time, the system itself is not committed to engaging in any intended self-study in action. The researcher may be researching patterns of statistical information or customer preferences, or writing a case history around a particular strategic initiative or a period of time. For the researcher, these are data which are gathered and analysed using established methodologies.

Alvesson (2003) addresses the situation of doing ethnographical research in one's own organization. He calls it 'self-ethnography' rather than insider ethnography or home culture ethnography. 'Self-ethnography' involves studying a setting in which the researcher is a participant on more or less equal terms with other participants. Participation comes first and is only occasionally complemented by research-focused observation, and hence 'observing participant' is a more appropriate term than participant-observer. The self-ethnographer relies on familiarity with the setting as the empirical starting point. If you are an insider researcher working within the ethnography tradition, you have to work hard at liberating yourself from subjectivity in order to meet the intellectual requirements of this research tradition. Young (1991) provides a good case of doing ethnography in his own organization. He reports how there was a culture of secrecy within the police force of which he was a member. Social science research was equated with 'claptrap' and anyone engaged in it ran the risk of being labelled a traitor and his or her promotion would be put in jeopardy. Therefore, he had to undertake his study surreptitiously and he refers to his work as 'writing espionage'.

The ethnographic role and the action researcher role are closely interconnected and sharply distinguished (Schein, 1987). The ethnographic observer attempts to be an unobtrusive observer of the inner life of an organization, while the action researcher works at enabling obtrusive change. Since the self-ethnography approach is not the subject of this book, we are not including any further exploration of this quadrant. In terms of the focus of this book, we are addressing research projects which are contained in quadrants 2, 3 and 4 of Figure 8.1.

Quadrant 2: classical action research

Quadrant 2 of Figure 8.1 applies where there is no intended self-study in action on the part of the researcher, while what is being studied is the system in action.

We see this as 'classical' action research, which comprises management action, internal consulting projects and some action learning. Coghlan (2003) refers to this as 'mechanistic-oriented' action research, by which he means that the research is framed in terms of managing change or solving a problem; it is directed at confronting and resolving a pre-identified issue.

There are many examples of quadrant 2 insider action research. Coghlan et al. (2004) provide a broad range of cases in which managers engaged in action research on particular issues confronting their companies: improving influence and performance in a multinational subsidiary, building interorganizational relationships in a global virtual community, implementing a 360-degree feedback programme and managing human resources, among others. Bartunek et al. (2000) also present three case examples of such research. Three of the authors, acting in their organizational managerial capacity, carried out action research projects in their own organizations – in a bank, a manufacturing company and a public utility. The research projects were oriented towards improving operational action, and there is no reported self-study in action on the part of the authors.

There is a broad range of accounts of quadrant 2 action research in nursing (Coghlan and Casey, 2001). Examples of action research in the clinical area illustrate interventions to improve: user participation in mental health nurse decision making (Tee et al., 2007), palliative care in generalist care settings (Hockley and Froggatt, 2006; Phillips et al., 2008), rehabilitation (Portillo, 2009), intensive care (Coyer et al., 2007) and care of older persons (Dewing, 2009). Action research in the organizational and operational areas include: regional framework for head injury care (Seeley and Urquhart, 2008), stroke service development (Jones et al., 2008) and community mental health service (Westhues et al., 2008).

Action research in quadrant 2 is typically the type of research undertaken in MBA and other master's programmes, where the focus is manager-led operational projects within a limited, specified time frame. These may well be projects already under way in an organization, and accordingly are opportunistically adopted by manager–MBA students as their action research project. Clearly, such opportunistic adoption is not confined to MBA situations, but may be undertaken by teachers, nurses, social workers, clinicians and similar people where course participants choose pre-existing projects for their own research topic. In these situations, action researchers attempt to bring the action research cycles of inquiry to a project which has not been set up as an action research project. This may make severe demands on action researchers' ability to manage organizational politics. Depending on the origin and scope of the project within this quadrant, the internal researcher may be working with an external consultant who has been hired to facilitate the change.

From their three case experiences of manager-led action research projects, Bartunek et al. (2000) generalize a number of relevant issues and themes:

1 The initial assignment to carry out work that leads to the action research project is likely to come from the manager's superiors and to be part of the manager's job description.
2 The other participants in the intervention are likely to be subordinates who need to buy into the change project.
3 The intervention is likely to be aimed at increased productivity.
4 Managers may find it helpful to constitute a consulting team to assist in the intervention.
5 Data gathering can take place through a variety of formal and informal means.
6 Feedback sessions can be integrated into the workday or conducted separately.
7 The manager is likely to have a personal stake in the outcome of the intervention.
8 The managers were all receiving training in action research while carrying out their interventions.

Quadrant 3: individual reflective study

Quadrant 3 of Figure 8.1 applies where the researcher is engaged in an intended self-study of herself in action, but the system is not. The researcher may be engaging in a study to improve professional practice; she is engaging simultaneously in a process of self-reflection, examining her own assumptions in action and learning about herself as events unfold. The researcher becomes a 'reflective practitioner' (Schon, 1983). Coghlan (2003) refers to this as 'organistic-oriented' action research, where the inquiry process is a value in itself. In organistic-oriented action research, researchers engage in an action inquiry process in which inquiry into their own assumptions and ways of thinking and acting is central to the research process. As mentioned earlier, Marshall (1999; 2001) describes this first person research/practice in terms of (1) inquiring into the inner and outer arcs of attention, (2) engaging in cycles of action and reflection and (3) being active and receptive. If you are locating yourself in this quadrant, then the processes described in Chapter 2 are paramount for you.

Krim (1988) provides a case example of quadrant 3 action research. In a somewhat dramatic account of an action inquiry project in action, he outlines the context of change in a city hall power culture, and describes both the political and conflictual dynamics within that culture and the processes of his own personal learning.

Krim (1988) reports how he, as initiator and coordinator of a new labour management cooperation programme based on employee participation in a city hall, sought to use himself as the key learning strategy, whereby his management style would be central to the inquiry process. He outlines the context of change in a city hall power culture, and describes both the political and conflictual dynamics within that culture and the processes of his own personal learning. He describes his reflection process in terms of a pyramid of five steps: recording and observing on a daily and hourly basis; a weekly selection and analysis of critical incidents; an exploration of these issues with his academic supervisor; rehearsal and role-playing with his supervisor in anticipation of further critical incidents; and a public testing in the real-life situation. He reports how this cycle of continuous rehearsal and performance allowed him to improve his performance in highly political and conflictual situations. From this process he received feedback on his management style, particularly on how he tended to 'de-authorize' himself, and so he adopted some practical rules of thumb to help him develop new behaviours. He reports how he was accused of spying when his research notes were pilfered from his computer and circulated among his antagonists.

You might read Krim's (1988) own account to have a fuller picture of his insider action research.

Questions for reflection and discussion

- What did it mean for Krim to 'use himself' as the key learning strategy? How does this example give you insights into quadrant 3 insider action research?
- If quadrant 3 work fits for you, how might you structure your own learning in action process?

A quadrant 3 research project may be pre-selected or it may emerge. Where the research agenda is self-selected by the researcher it may focus on the researcher's job or role within the organization.

It may be that the quadrant 3 research emerges out of a quadrant 2 project. Meehan and Coghlan (2004) describe a case in which Meehan, in his organizational role, was asked by his superior to evaluate the addiction counselling service in his regional health area. The service was primarily treatment focused and provided a service to help those with addiction problems end their dependence on mood-altering substances and to rebuild their relationships with their families, friends and colleagues.

Meehan met the group of counsellors. As the group explored the issue of evaluation, resistance emerged. There was concern that, no matter how good the evaluation was, the organization would not accept it. The alienation that the group felt was surfacing, and an air of futility about the process was being articulated. At this time, Meehan found himself actively listening to the group and he speculated that, if an effective evaluation was to take place, the feelings of the participants needed to be heard and dealt with in some way. He was becoming aware that he was utilizing listening and counselling skills in the process. He had learned counselling skills as a clinical psychiatric nurse, and they enabled the group to explore feelings they had about these issues. As time went on he became more convinced that he was working on a project that would really benefit from an action research approach and he entered into discussions with the group on the issue. His thinking in this regard was that the action research approach would be likely to produce insights that could not be gleaned in other ways. He hoped that the process would enable the group to get new and innovative insights into the service and to take action based on them.

Meehan introduced the group to the notion of cooperative inquiry, and the group agreed to pursue the issues pertaining to themselves and the service in that mode of reflection on experience through a cooperative inquiry group process. Over a six month period the group met 11 times and explored three themes: feelings of alienation and powerlessness, the lack of strategic direction in the service, and professional identity and autonomy. They carried out a SWOT (Strengths, Weaknesses, Opportunities and Threats) analysis of the service and identified areas for change, both within themselves and in the management of the service.

As he led the group through the formulation of an action plan to initiate a service evaluation, Meehan found himself confronted by a feeling of futility within the group. On one level he was getting feedback that his intervention was helpful, but how it was helpful was not articulated. On another level the group was sceptical about the usefulness of the evaluation. He assumed that there was a feeling it was all done before and 'What's the point?' He then began to see his role as a go-between or an arbitrator between the management team and the addiction counsellors. In some way he perceived that if he could get a good evaluation of the process utilizing management input, addiction counselling input and from a service users' perspective, that would solve the issue. However, intuitively he was picking up resistance and vibes saying that this is not going to work.

The group of addiction counsellors worked to reflect on their experience and to develop new and creative ways of looking at things. They also learned how to act to change things they wanted to change and explored how to do things better. Each member acted as a co-subject in the reflection phases and a co-researcher in the action phases. They enacted the action research cycles of reflection and action in a psychologically safe environment, which enabled them to make sense of their experience and take steps to initiate change. Overall the outcomes of the cooperative inquiry process were that there was a reduction in the feelings of alienation and powerlessness in the group, and this was evidenced in the enthusiasm with which the group subsequently engaged in the evaluation. There grew a commitment to work in partnership with all stakeholders and service users in evaluation and in developing the service. The group was beginning to look at issues from other perspectives and this appeared to reduce hostility and fear. There were also real issues about their power in the organization: at one level they felt like victims in the process, yet it was obvious that unless they agreed with changes they were unlikely to get implemented.

Meehan found the whole process difficult yet rewarding. It was an ongoing dynamic and he found he learned a lot in the process. It was the first time he had consciously bridged the gap between his roles as a clinician and as a manager. It showed him how to utilize his skills as a clinician in management. He found the action research cycles to be very useful and they produced insights which he believed could not be gleaned in other ways. The process was also a healing one and he believed it empowered the group to deal with underlying difficulties in the service.

Through this process Meehan believed he received insights into how he could further develop as a manager and a person. One particular change in his behaviour and role as a manager that occurred in this process was how he now perceived his role differently. Traditionally, he had viewed his role as an agent of control; now he viewed himself in a different way. The process enabled him to explore new and interesting ways of viewing his role as a manager; it gave him an insight into potential ways of working with alienation and powerlessness in the workplace. Hitherto he had functioned as what Frost and Robinson (1999) describe as a 'toxic handler', which he now viewed as an unhealthy way of managing organizational pain. He now saw one of his managerial roles as that of healer (Quick et al., 1996).

In hindsight he was able to manage many of the political elements because as an insider he understood the process politics of the organization. He found himself as a middle manager who has to understand internal and external pressures on the organization and satisfy the personal or competing interests. The personal and emotional issues remained confidential to the group but the broader learning from the group was discussed. Changes to the management structures, issues regarding supervision, renegotiations regarding time frames for completion were all dealt with well and there was no conflict on these issues.

You might read Meehan and Coghlan (2004) to obtain a fuller picture of Meehan's insider action research project.

Questions for reflection and discussion

- How did Meehan's work in quadrant 2 lead to work in quadrant 3? Did he leave quadrant 2 behind in his pursuit of quadrant 3 learning?
- If you are beginning in a quadrant 2 setting, what possibilities and options are there for you to move into quadrant 3? Do you want to move to quadrant 3 or do you prefer to remain in quadrant 2? Is being in both possible, as it was for Meehan?

Quadrant 4: transformational change

Quadrant 4 of Figure 8.1 is where both the researcher and the system are engaged in intended study in action. The system has made or is making a commitment to change. For example, the system may undertake a system-wide transformational change programme in which everything in the system is open to review, as instanced in the movement to quality of working life (QWL), business process reengineering and organization development, or through a learning history approach. In this instance, there is a broad commitment to reflecting on experience and learning. The researcher's role involves being part of this collective reflection and learning and articulating what is happening. There is active participation by both the system and the individual. In a large scale system change project, it is likely that there are external consultants: hence the need for insider–outsider collaboration (Bartunek and Louis, 1996; de Guerre, 2002; Adler et al., 2004; Shani et al., 2008). Quadrant 4 is clearly the most difficult and demanding in terms of scale, complexity, and conceptual, analytic and practical knowledge and skill. We are not providing an example but we suggest you engage with some of the accounts of learning history (Kleiner and Roth, 2000; Roth and Kleiner, 2000).

The research process itself may provoke a move from one quadrant to another. An action researcher operating in quadrant 2, for instance, may find that the technical problem being researched is a symptom of underlying cultural assumptions, and so its resolution carries more far-reaching implications than was envisaged at the outset. The dimensions of research in quadrant 3 may evolve into quadrant 4. It may be that the researcher's personal development through the research process involves a gradual movement from quadrant 2 to quadrant 3. Participants in the master's programme in management practice at the Irish Management Institute, University of Dublin, which is a part-time action research oriented programme, reported that they perceived the two year programme as a journey from quadrant 2 to quadrant 3, with quadrant 4 the desired outcome in the long term.

Eddie reflects on his experience:

> I would have to say that it began in quadrant 3. I, as a classroom teacher, was engaged in an intentional self-study in action to gain insight with respect to the inquiry questions of how the learning of beginning algebra can be made more meaningful to students, especially those who have faced prior difficulties with mathematics. However, the larger system that I found myself in was not engaging in this type of inquiry. As a result the project took the form of me, as an individual, engaging in a reflective study of my own professional practice.
>
> As the project ran its course, however, I feel as though it moved into quadrant 4. I followed my supervisor's suggestions and engaged my fellow teachers in the analysis of the Foundations of Algebra I curriculum. After writing it, I gave it to them and asked them in the spirit of co-inquiry whether or not what I had prepared connected to their reality. I was especially interested in what they would add and how they would describe the same things. As a consequence I witnessed my project gain momentum and buy-in. Not only were my colleagues interested in gaining an understanding about the question that was at the heart of my project, but the school's administration also reacted by expressing interest in the work and a willingness to provide any necessary supports. For these reasons I am excited to say that the system eventually participated with me in an intended study in action. The result of this will hopefully be a large scale transformational change in the way in which introductory algebra is taught at my school.

George reflects:

> Much of the project would be placed in quadrant 3, since I was engaging in an intended self-study of myself in action, but my department was not. Since my department was pursuing a year-long goal, many of their actions could be characterized as quadrant 2, internal consulting and action learning. In the middle of launching the initiative I became our department's team leader, so in the process of developing the agenda for our weekly meetings and reporting on the progress made toward our goals, the administration and the teachers began to participate in the cycle of observing, planning, implementing and reflection. This occurrence had some characteristics of quadrant 4, whereby the researcher and the system are engaging in an intended self-study. While we never explicitly articulated the engagement in self-study, this is what we were doing. This understanding leads me to realize that you may be able to motivate people more readily to develop professionally by appealing to their own curiosities in reflecting on goals that they feel are beneficial to their practice. Dropping the term 'project' removed a psychological burden. To many people the word 'project' means a 'lot of hard work'. It wasn't necessary to define this initiative as a project in order for the teachers to engage.

The grid acts as a mechanism for subject selection. You may select a subject based on (1) a desired outcome for yourself and/or your system, (2) the opportunity or access to areas of the organization, and (3) your sense of your possession of the level of skill required to work in any particular quadrant.

Action research at home

Not all action research in your own organization is done at places of work. The following example illustrates action research at home.

Goode and Bartunek (1990) described a self-initiated action research project in an apartment complex, where Goode initiated a process to address a problem of direct personal concern. In the apartment complex where Goode was a resident, there were problems relating to mail delivery and security. Single efforts to address resident concerns had been unsuccessful. Goode approached several concerned residents, discussed the problem with them and explained action research in order to foster an environment which enable collaboration and search for multiple causes of the problem, rather than the unitary blaming of the complex's caretakers, which was common. Goode then formulated a plan for gathering information, which involved: (1) producing an informational letter which she distributed to all residents, (2) conducting preliminary meetings with some of the residents and the caretakers, (3) conducting an informational session with anyone interested in participating, and (4) interviewing all interested individuals. All these actions took place and the data generated were analysed under such headings as: organization strengths, existing mail security problems, and existing group-level structures and problems. A feedback session was held at which task forces were formed to address particular issues and actions were implemented.

Goode and Bartunek reflect on this case in terms of two issues. First, this case is an example of action research in an underbounded system. As this subject is not central to the theme of this book, we will not reflect further on this point. Second, and which concerns us in the context of action research in your own organization, is how the action researcher initiated research in an organization where she had a personal stake. Goode and Bartunek point to two roles played by the action researcher. One role was that of a long-term participant in the system, which meant she shared the concerns of the fellow residents and had credibility. The other role was a short-term consultant role where her knowledge of action research provided guidance for the process. She was willing to take an educative, directive and participative approach to enable the resolution of problems and the emergence of new structures.

Questions for reflection and discussion

- What insights are you receiving from this informal action research initiative?
- Are there opportunities for you to practise your insider action research skills in situations outside the dissertation setting?

Conclusions

In this chapter we have reflected on the subject of doing research in and on your own organization. This involves undertaking research while a complete member, which in this context means wanting to remain a member within your desired career path when the research is completed. In undertaking action research in and on your own organization, the commitment to learning in action by both the system and yourself is a useful defining construct. If the research project is accompanied by the system's commitment to learning in action, then secondary access is easier. If your organizational role is that of internal consultant, then your research role is integral to that role. If on the other hand you are a manager, then you are taking on the researcher role in addition, and that may create confusion.

As an insider action researcher you are engaged in first person research, using your preunderstanding of organizational knowledge and organizational studies for your own personal and professional development. You are engaging in second person research by working on practical issues of concern to your organization in collaboration with colleagues and relevant others. You are engaging in third person research by generating understanding and theory which are extrapolated from the experience.

Doing action research in your own organization involves (1) clarifying the action research project in terms of both your and the system's commitment to learning in action, and (2) managing issues of role and secondary access.

Exercise 8.1　Assessing your research focus

Drawing on Figure 8.1, reflect on your current position.

- What quadrant are you in?
- What quadrant do you think that you are in?
- Which quadrant are you best at?
- Which quadrant would best serve your dissertation requirements?

NINE

Preunderstanding, Role Duality and Access

The dynamics of doing action research in your own organization in quadrants 2, 3 and 4 (Figure 8.1) involve building on the insider knowledge you already have (preunderstanding), managing the two roles you have (your standard organizational roles and now, in addition, the action researcher role) and negotiating access. The questions underpinning this chapter are: how do you build on the closeness that you have to the organization and yet maintain distance? How do you balance the potential dilemmas and tugs between your established organizational roles and your researcher role? We now explore these.

Preunderstanding

'Preunderstanding refers to such things as people's knowledge, insights and experience before they engage in a research programme' (Gummesson, 2000: 57). Knowledge refers to what is generated during the research. The knowledge, insights and experience of the insider researcher apply, not only to the theoretical understanding of organizational dynamics, but also to the lived experience of your own organization. Some misunderstand the notion of preunderstanding and equate it with tacit knowledge. Preunderstanding includes both explicit and tacit knowledge. Personal experience and knowledge of your own system and job comprise a distinctive preunderstanding for the insider researcher.

One advantage you have as an insider researcher over an outsider researcher is that you have valuable knowledge about the cultures and informal structures of your organization. Organizations lead two lives. The formal or public life is presented in terms of its formal documentation: mission statement, goals, assets, resources, annual reports, organizational chart, and so on. The informal or private life is experiential, that is, it is the life as experienced by its members: its cultures, norms, traditions, power blocs and so on. In their informal lives, organizations are centres of love, hate, envy, jealousy, goodwill and illwill, politics, infighting, cliques, political factions and so on – a stark contrast to the formal rational image that organizations tend to portray. You have valuable experience of this, though you don't know it all. While this knowledge is an advantage, it is also a disadvantage as you are likely to be part of the organization's culture and find it difficult to stand back from it in order to assess and critique it. You may need to be in tune with your own feelings as an organizational member: where your feelings of goodwill are directed, where your frustrations are and so on.

Nielsen and Repstad (1993) outline some examples of such experience and preunderstanding. You have knowledge of your organization's everyday life. You know the everyday jargon. You know the legitimate and taboo phenomena of what can be talked about and what cannot. You know what occupies colleagues' minds. You know how the informal organization works and whom to turn to for information and gossip. You know the critical events and what they mean within the organization. You are able to see beyond objectives which are merely window dressing. When you are inquiring you can use the internal jargon and draw on your own experience in asking questions and interviewing, and you can follow up on replies and so obtain richer data. You are able to participate in discussions or merely observe what is going on without others being necessarily aware of your presence. You can participate freely, without drawing attention to yourself and creating suspicion.

There are also some disadvantages to being close to the data. When you are interviewing you may assume too much and so not probe as much as if you were an outsider or ignorant of the situation. You may think you know the answer and not expose your current thinking to alternative reframing. In insider research, epistemic reflexivity is the constant analysis of your lived experience as well as your own theoretical and methodological presuppositions. This helps you to retain an awareness of the importance of other people's definitions and understandings. You may find it difficult to obtain relevant data because as a member you have to cross departmental, functional or hierarchical lines, or because as an insider you may be denied deeper access which might not be denied to an outsider. Ferguson and Ferguson (2001) caution against insider action researchers believing that they fully understand their own contexts, when in fact their perspectives are only partial. The moral in their view is to be honest about the perspectives from which you operate and be open to disconfirming evidence – perhaps seeking it out through interviews.

Table 9.1 outlines the essential tasks and skills required to work with the strengths and limitations of preunderstanding. The task at first person is to develop a spirit of inquiry so as to receive insights into familiar situations where things are taken for granted because they are so familiar. This requires self-awareness reflection skills (as we explored in Chapter 2) so that you learn to attend to and to question your own assumptions. At second person, the challenge is to develop

Table 9.1 Preunderstanding in first, second third person practice

	Task	Process
First person	Develop a spirit of inquiry in familiar situations where things are taken for granted	Attending to and questioning own assumptions; self-awareness reflection skills
Second person	Develop collaborative inquiry and action in familiar situations where the spirit of inquiry may be diminished	Collaborative inquiry and action: • combine advocacy with inquiry • have range of intervention options • test assumptions and inferences • we learning window
Third person	Develop practical knowledge of how to inquire as a 'native'	Link practice with theory

collaborative inquiry and action with relevant colleagues in familiar situations where the spirit of inquiry may be diminished. So you need to be able to combine advocacy with inquiry, test assumptions and inferences through having a range of intervention options, and be skilled at using each of them appropriately. The learning window is a useful framework for both first and second person work on preunderstanding. Your third person contribution may include an understanding of developing practical knowledge of how to inquire as a 'native' and so to be able to link theory with practice.

Schein (2004) describes organizational culture as patterns of basic assumptions which have been passed on through generations of organizational members and which are unnoticed and taken for granted. Accordingly, the approach to uncovering cultural assumptions is a dialogue between organizational members and an external process consultant who facilitates the exploration of what assumptions underlie artefacts and values. As an insider researcher you may need an external facilitator to help you make sense of your experience. The academic supervisor may play this role. Krim (1988) reported how, in his meetings with his academic supervisor, he would roleplay critical incidents. These role-plays were important in his reflective learning process.

Gorinski and Ferguson (1997) share a conversation about their respective experiences of being an insider action researcher. They identify positive aspects of accessibility, credibility, trustworthiness, commitment and familiarity with the research context and personnel. They found the barriers to be: communication difficulties, time limitations, power positions and how cultural openness changed as they found themselves being left out of things. Their conclusion is that they had to 'make the road by walking'.

The quest for authenticity, with respect to preunderstanding, requires attention to your own inner and outer dynamics. The outer dynamics pertain to the familiarity of the settings and inquiry as to how they may be different and what further insights are required, where the ritual appears to be the same. You may be different this week because of what you've thought, felt and done or because of what has been done to you. Likewise the other participants. The context may have shifted slightly or subtly. The inner dynamics point to your thoughts and feelings in such familiar settings, which may be familiar feelings of powerlessness or other feelings or thoughts that accompany being in these regular situations with the same people. Being intelligent refers to asking what evidence you are being presented with as you work in a familiar setting and what it is that you take for granted. Being reasonable means weighing the evidence and distinguishing inferences and attributions in deciding what conditions are fulfilled in order to make a confident judgement. Being responsible means that you take responsibility for what you decide to do and what you do.

Journaling is an important mechanism for learning to reflect on and gain insights into your preunderstanding. Through recording your experiences, thoughts and feelings over time as you move through your project, you can begin to identify gaps between what you think you know and then find that you don't, or between explicit and tacit knowledge. You can begin to learn to stand back and critique what you have taken for granted hitherto. The journaling process helps your meta learning of content, process and premise in the arena of your preunderstanding, and the integration of what you know already because of your closeness to the issues and the organization with what you are discovering as you engage in first, second and third person inquiry.

Role duality: organizational and researcher roles

In action research the traditional distinction between the researcher and the researched diminishes. Working in quadrants 2, 3 and 4 of Figure 8.1, the insider action researcher role is added to the complete member role, discussed earlier.

In a detailed and insightful reflection on role duality, Wirth (Ravitch and Wirth, 2007) reflects that negotiating between her various roles and identities was the greatest challenge, as it was difficult to develop a balance between being a colleague, a school teacher, a friend and a researcher. In one role she acted as a leader supporting colleagues in a literacy programme and ensuring that she did not act as an evaluator. In her researcher role she needed to be more direct in presenting the expectations of the research initiative. She notes that what was important was that she was an insider, and that as colleagues knew her and she knew them, they seemed to assume fairly that she had no hidden agenda and that the more sensitive issues involving culture, race, social class and equity could be approached. Because she knew and respected them she understood the suggestions they made about the design and implementation of the initiative. So she felt able to recognize their concerns and work to address them. At the same time, there were some significant challenges and disadvantages. She wondered that, because they knew her, some seemed to think they could ignore her requests.

She reports that one of her most significant struggles was to maintain a collaborative mode of working with the teachers in a school climate that did not foster collaboration. Being familiar with the various personalities, perspectives and dynamics between the teachers was helpful in setting up the collaborative effort and facilitating change, but she noted that in other ways her preconceived notions about the teachers compromised her ability to see them more objectively and constrained possibilities for building new kinds of relationships and processes.

Questions for reflection and discussion

- Do you recognize the challenges that Wirth presents? We suggest that you read the complete article (Ravitch and Wirth, 2007) and that you reflect on for yourself and discuss with your group the issues that she identifies and how she dealt with them.
- What roles do you play in your organization? When you add the researcher role, what challenges do you expect to face or are facing?

Williander and Styhre (2006) reflect on the role of the insider action researcher as a dual channel between academia and practice, i.e. information flows in both directions. In their experience, the first role of the insider action researcher was to act as a bridge between academia and the company and in translating information between the two worlds of research and practice. The second role was that of traditional researcher, but using the benefits of being an insider, through constructing case studies about the company and its parent company. Studies on one and the same case from a network perspective, a marketing perspective, a consumer behaviour perspective and so on provided multidisciplinary understanding of the issue and became an important foundation for the emerging ideas and how to address them. The third role of the insider action researcher was to use the company as an object of experimentation. Academia would have the opportunity of testing, observing, intervening and validating, while the company would gain competitive advantage from the first hand knowledge created.

We suggest you read Williander and Styhre (2006) for a fuller exploration of their reflection on the role of the insider action researcher.

Ashforth et al. (2000) explore the nature of roles. They present a number of useful constructs. Role boundary is defined as the scope of a role. Role boundaries can be flexible (that is their boundaries can be pliable, spatially and temporally) and they can be permeable. (You may be physically in one role and psychologically and/or behaviourally in another.) For Ashforth et al. these constructs of role flexibility and permeability enable transition from one role to another. In terms of doing action research in your own organization, you may be in your office or at a meeting in your organization exercising your organizational role (physical and spatial) and at the same time probing for answers to questions in your research role.

Ashforth et al. also explore the notion of role identity, that is, how a role cues specific goals, values, behaviour and so on. Role identities are socially constructed definitions of self-in-role, consisting of contrasts between core and peripheral features. In their view, combining the construct of role boundary in terms of flexibility and permeability with that of role identity in terms of contrast between core and peripheral features can be arranged along a continuum from high role segmentation to high role integration.

Roth et al. (2004) apply Ashforth et al.'s framework to three case examples in which insider action researchers spanned boundaries as they enacted roles in both the organizational setting and the academic setting. For example, they applied frames of reference from the academic setting to the organizational setting. They note that each of them acted as a bridge between industry and the academy which could facilitate the joint knowledge creation necessary for both settings, albeit in a different way in each project. In their reflection, they considered that the nature of the insider action researcher's dual role facilitated the creation of a stable bridge between the academy and industry through extensive knowledge of the company being studied and the research process.

Augmenting your normal organizational membership roles with the research enterprise can be difficult and awkward, and can become confusing and overwhelming. As a result, in trying to sustain a full organizational membership role and the research perspective simultaneously, you are likely to encounter role conflict. Your organizational role may demand total involvement and active commitment, while the research role may demand a more detached, theoretic, objective and neutral observer position. This conflict may lead you to experience role detachment, where you begin to feel as an outsider in both roles (Adler and Adler, 1987).

Humphrey (2007) reflects that her insider-outsider status was a valuable resource for her as she could mobilize both insider wisdom and outsider research to explain the rationale of self-organization. Some expressed relief that issues had been exposed by an 'insider' rather than an 'outsider' and so they were now 'on the table' for debate and that there were 'theories' to make use of them. She reflects:

The perils of the insider-outsider are that she can be pushed and pulled along an invisible insider-outsider continuum by others who have a vested interest in who she is and what she is doing, and she is cast now as an 'insider' and now as an 'outsider' by different actors and audiences and can lose her sense of herself. To actively take charge of the hyphen is to appreciate one's uniqueness as an insider-outsider and to cultivate the art of crossing between life-worlds. (2007: 23)

Questions for reflection and discussion

- Do you recognize the perils that Humphrey (2007) presents? We suggest that you read the complete article and that you reflect on for yourself and discuss with your group the issues that she identifies and how she dealt with them.
- What pulls and tugs are you experiencing in your insider action research project? Are they pushing you to choose one role identity over another? How do you deal with them?

Your organizational relationships are typically lodged and enmeshed in a network of membership affiliations, as you have been and continue to be a participant in the organization. These friendships and research ties can vary in character from openness to restrictiveness. You are likely to find that your associations with various individuals or groups in the setting will influence your relations with others whom you encounter, affecting the character of the data you can gather from them.

Table 9.2 outlines the essential tasks and skills required to work with the challenges of role duality. The task at first person is to hold and value both sets of roles simultaneously and learn to catch your internal responses to conflicting demands and to deal with them. This challenge may get played out in second person where you try to hold and manage the demands of both roles, particularly in situations of conflicting role demands. In such situations your ability to negotiate your role with significant others will be

Table 9.2 Role duality in first, second and third person practice

	Task	Process
First person	Holding and valuing both sets of roles simultaneously	Catching internal responses to conflicting demands and dealing with them
Second person	Holding and managing demands of both roles, particularly in situations of conflicting role demands	Role negotiation with significant others
Third person	Develop practical knowledge of how dual roles impact on action research and contribute to insider action research role, identity theory	Linking experience of role duality with theory

important. At third person you will have developed practical knowledge of how dual roles impact on action research and so you may contribute to insider action research role identity theory.

Your organizational role or roles will influence the degree of role confusion or ambiguity that you will experience as an insider action researcher and your ability to cope with your situation. If your sole job in your organization is that of internal change consultant, then you are already a researcher in your own organization. We see this as a single role with low potential for role confusion. Quadrants 2 and 4 of Figure 8.1 incorporate such internal researchers. On the other hand, if your job is that of manager, then you are taking on a researcher's role in addition to your managerial job. Hence you have to manage dual roles, manager and researcher, and there is high potential for role confusion. Quadrant 3 of Figure 8.1 is a good example of this situation. Single role researchers whose job involves research type activities are quite distinct from dual role researchers whose research activity is a separate role from their standard functional role. In terms of the cases we discuss throughout the book, we can see that Krim (1988) and Moore (2007) are examples of dual role researchers, while Humphrey (2007) is a single role researcher.

Feelings of detachment can be oriented toward one or other of these roles and increase or decrease as the research progresses. When you are caught between the loyalty tugs, behavioural claims and identification dilemmas you initially align yourself with your organizational role. Elizur (1999) uses the term 'self-differentiation' to review how he, as an insider consultant, managed to (1) contain emotions and relate to emotionally charged issues in a balanced way, and (2) maintain his own autonomy and self-identity in these situations. Your involvement in the two roles affects your relations with organizational members (Adler and Adler, 1987). The new dimension of your relationship to members and/or your new outside interests sets you apart from ordinary members.

Roth et al. (2007), reflecting on their insider action research work, identify challenges with regard to multiple roles. They noted that the insider action researcher has three roles: he or she 'is first an employee of his/her organization and second is a researcher. In addition to these two roles a third role evolved when the insider action researcher was seen as an internal consultant to the organization' (2007: 51). They identified four challenges:

- operating successfully within each role
- translating ideas into the appropriate language for each world
- knowing the language of the organization and the academic community, which is important for reporting back data from relevant interventions in the proper context
- learning to 'put on different hats' so that multiple roles are an asset rather than a burden.

- In what roles are you finding yourself?
- How are you crossing the boundaries between the organization and the scholarly community and translating language from one community to the other?
- How are your multiple roles an asset rather than a burden?

Nielsen and Repstad (1993) cite a number of specific role duality related advantages and disadvantages of insider research. You may have a strong desire to influence and to change the organization. You may feel empathy for your colleagues and so be motivated to keep up the endeavour. These factors are beneficial in that they may sustain your energy and be a drawback in that they may lead to erroneous conclusions. You have to deal with the dilemma of writing a thorough report and dealing with the aftermath with superiors and colleagues on the one hand, and doctoring your report to keep your job on the other. When you are observing colleagues at work and recording your observations, you may be perceived as spying or breaking peer norms. Probably the most crucial dilemma for you as an action researcher in your own organization, particularly when you want to remain and progress in the organization, is managing organizational politics. We will return to this subject in Chapter 10.

Roth et al. (2004) recount how a conflict between the two roles arose in terms of timing. In the organizational setting there was pressure to present conclusions from the research setting, even though in research terms the insider action researcher thought it was too early and the data had not been fully reflected on. In their view, 'In doing so the insider action researcher tried to legitimize the researcher role before it was mature in that setting, giving the opposite result' (2004: 212).

A practical issue you have to deal with is that you may be too close to the issues and the people in the organization and so you have to work more consciously and explicitly at the process of inquiry. If you have been trained in a particular discipline or are familiar with a particular function you may not be open to seeing problems from other perspectives. You may be too close to the people and the situations you are researching. What if the research involves critiquing your friends or close colleagues? It may be difficult for you to stand back from the situation and question your own assumptions, which heretofore have been unquestioned. Epistemic reflexivity, correctly enacted, incorporates challenging presuppositions derived from closeness to people and issues.

Homa (1998) reflected on what it was like to combine the roles of CEO and researcher. He provided a number of useful pieces of potential advice:

1 You need to be reasonably on top of your job as it is hard to switch psychologically from management responsibility to research without it. Therefore, selecting the right time in your

career to do research is an important choice. You need to achieve effective personal organization, good time management and the ability to create a distance between work and study so that you can leave the organization for periods of uninterrupted study.

2   You need excellent secretarial support, particularly if you don't do your own typing.

3   Over time you need to balance the achievement of being a manager and working through others with the solitary work of a researcher.

4   You need a strong management team and a strong and supportive chairman.

The quest for authenticity with respect to role duality requires attention to the inner and outer dynamics of holding both sets of roles. The outer dynamics pertain to the roles expected of you and those you seek to adopt and the ambiguity or conflict between them. The inner dynamics refer to how you think and feel about any such role ambiguity or conflict and the implications that you internalize. Being intelligent refers to understanding the structural nature of roles and how they are played out in this organizational setting. Being reasonable means weighing the evidence and judging the demands of conflicting role expectations. Being responsible means you take responsibility for what you decide to do and what you do.

Access

Primary access refers to the ability to get into the organization and to be allowed to undertake research. So as you are already a member of the organization, you have primary access; you are already in. While you have primary access, you may or may not have secondary access, that is, you may or may not have access to specific parts of the organization which are relevant to your research. This is especially true of research in quadrant 3 where the system is not committed to self-study in action. By parts of the organization we mean not only functional areas such as departments, but also hierarchical areas whereby there is restricted access to specific privileged information, which may not be available otherwise. Insider researchers do find, however, that membership of the organization means that some avenues are closed to them because of their position in the organization. Clearly, any researcher's status in the organization has an impact on access. Access at one level may automatically lead to limits or access at other levels. The higher the status of the researcher, the more access she has or the more networks she can access, particularly downward through the hierarchy. Of course, being in a high hierarchical position may exclude access to many informal and grapevine networks. Fundamentally, secondary access means access to documentation, data, people and meetings. In relation to research in quadrants 2 and 4, the system takes responsibility for secondary access because it is committed to self-study in action.

An important aspect of negotiating the research project is to assess the degree of secondary access which one is allowed. Dual role researchers may experience more problems than single role researchers. Of course, what is espoused at the outset and then actually allowed may be different once the project is under way and at a critical stage. There may be a significant gap between the aspiration towards 'purity' of research and the reality. How access is realized may depend on the type of research being undertaken and the way information is disseminated.

Negotiating access with your superiors is a tricky business, particularly if the research project aims at good work and not something bland. It raises questions about the different needs which must be satisfied through the project. As an insider action researcher you have needs around doing a solid piece of research which will contribute to your own career and development (first person research, for me). You also have needs around doing a piece of research in the organization which will be of benefit to the organization (second person research, for us) and contribute to general theory for the broader academic community (third person research, for them). Balancing these three audiences is difficult. In general, researchers' superiors have needs around confidentiality, sensitivity to others and organizational politics.

For you, the researcher, who are undertaking research as part of a degree programme or who seeks to publish, a particular issue relating to access is the fact that what is researched will be going outside the organization. Theses and dissertations are read by people external to the organization and are filed in libraries, with their abstracts disseminated to a wider audience. Bartunek and Louis (1996) see this as an 'outsider' role that the insider also plays. At the extreme, organizations can be paranoid about information going outside the organization, or at least be nervous about it.

Once again we emphasize the value of journaling in exploring role duality. Role ambiguity and role conflict can challenge how your role identity is flexible and permeable. Locating yourself on the continuum between role integration and role segmentation and exploring the forces whereby you are enabled or inhibited in exercising both your organizational roles and your insider action researcher role are key to first and second person inquiry and practice.

Conclusions

Role duality and secondary access tend to be research project specific and organizationally dependent, whereas preunderstanding tends to be researcher specific. Therefore, preunderstanding is not directly linked to the quadrant schema presented in Figure 8.1. Preunderstanding, as the word suggests, is what the insider brings to the research process. In summary, secondary access pertains to the specific nature of the research project. In quadrants 2 and 4 the system takes responsibility for secondary access, and it may be more readily available if all the relevant parts of the system are committed to the project. It is more problematic in quadrant 3, where the system does not necessarily have a commitment to your action research.

Exercise 9.1 Assessing your preunderstanding

Apply the learning window (see Exercise 6.3) to the area of your own preunderstanding and understanding.

1 Take a particular issue within your insider action research project.
2 Fill in the boxes, distinguishing between what you know you know; what you think you know and so need to test your assumptions, inferences and attributions; what you know you don't know and so need to inquire more in order to find out; and where you need to be open to new areas of discovery.
3 Apply this process to other specific issues.

1 What I know	2 What I think I know
3 What I know I don't know	4 What I don't know that I don't know

TEN

Managing Organizational Politics and Ethics

While doing any research in an organization is very political (Punch, 1994), doing research in and on your own organization is particularly so. The questions under-pinning this chapter are: how do I survive and thrive in a political environment? How do I act politically in a mode within the ethics of action research? In this chapter we examine the politics and ethics involved in doing action research in your own organization. Power is a subject that is well explored from multiple per-spectives and in various literatures. It is not our intention in this book to retrace the research and discussion of power and politics in society and organizations. We do, however, recommend strongly that you undertake some reading in this area so as to ground your own first, second and third person inquiry and practice. Buchanan and Badham (2008) provide an excellent exploration of the subject of power and politics which we think you, as an insider action researcher, will find very useful and which we hope will facilitate insights.

As Table 10.1 illustrates, there are tasks and processes for first, second and third person practice. In first person terms, the task is to learn to act politi-cally and authentically in a mode within your own ethics and those of action

Table 10.1 Politics and ethics in first, second and third person practice

	Task	Process
First person	Learning to act politically in a mode within the ethics of action research	Acting politically and authentically
Second person	Surviving and thriving political dynamics	Performing and backstaging
Third person	Articulating knowledge out of action that is actionable politically; contributing knowledge of what organizations are really like	Linking political experience with theory

research. In second person terms, the task is to survive and to thrive in the political dynamics in interaction with others. A third person contribution is to articulate actionable knowledge out of political action.

The politics of researching your own organization

Clearly any form of research in any organization has its political dynamics. Political forces can undermine research endeavours and block planned change. Gaining access, using data, disseminating and publishing reports are intensely political acts. Take for example the act of framing, which we discussed in Chapters 3 and 7. Framing is never a neutral act; it rarely affects all stakeholders in the same way. Some may benefit, whereas some may feel threatened and may feel that they will be harmed because framing exposes weaknesses in performance. So while in action research framing is a collaborative activity, raising certain questions and applying judgements to particular issues may have severe political implications.

Therefore, doing action research in your own organization is political. Indeed, it might be considered subversive. Action research has a subversive quality about it. It examines everything. It stresses listening. It emphasizes questioning. It fosters courage. It incites action. It abets reflection, and it endorses democratic participation. Any or all of these characteristics may be threatening to existing organizational norms, particularly in those organizations that lean towards a hierarchical control culture. Goffman (1959) refers to insider roles as 'informer', 'confidant' and 'renegade'. Clearly you do not want to be viewed as a 'renegade'. Meyerson (2001) calls those who quietly enact change in their own organizations as 'tempered radicals'. Cooklin (1999) refers to the insider change agent as the 'irreverent inmate', one who is a supporter of the people in the organization, a saboteur of the organization's rituals and a questioner of some of its beliefs. While as the action researcher you may see yourself as attempting to generate valid and useful information in order to facilitate free and informed choice so that there will be commitment to those choices in accordance with the theory and practice of action research (Argyris and Schon, 1996), you may find that, as Kakabadse (1991) argues, what constitutes valid information is intensely political.

Accordingly, you need to be politically astute in engaging in action research, becoming what Buchanan and Badham (2008) call a 'political entrepreneur'. In their view, this role implies a behaviour repertoire of political strategies and tactics and a reflective self-critical perspective on how those political behaviours may be deployed, and emphasizes the risk-taking and creative role of the change agent. Buchanan and Boddy (1992) describe the management of the political role in terms of two activities, *performing* and *backstaging*. *Performing* involves you in the public performance role of being active in the change process, building participation for change, pursuing the change agenda rationally and logically, while backstage activity involves the recruitment and maintenance of support and the reduction of resistance. *Backstaging* comprises skills at intervening in the political and cultural systems, through justifying, influencing and negotiating, defeating opposition and so on. Because you are an insider you have a preunderstanding of the organization's power structures and politics, and are able to work in ways that are in keeping with the political conditions without compromising the project or your own career.

As you engage in your action research project, you need to be prepared to work the political system, which involves balancing the organization's formal justification of what it wants in the project with your tacit personal justification for political activity. Throughout the project you will have to maintain your credibility as an effective driver of change and as an astute political player. The key to this is assessing the power and interests of relevant stakeholders in relation to aspects of the project. One particular manager may have a great deal of influence with regard to strategic decision making, but little influence with regard to budget allocation.

Pettigrew (2003) reflects on his own role as a political entrepreneur. He notes that it can be exhilarating when it appears that your advocacy, enthusiasm and energy have created desired effects towards some defined outcomes and equal and opposite despair when things go wrong. He reflects that there's a fine line between acting in a politically astute manner and acting unethically. In his view, action researchers have to build relationships and trust with people who operate from different mental models and at different levels. Yet working as a change agent cannot always be done with openness, honesty and transparency. He judges that the real skill for the political entrepreneur is knowing that the game is everything and that it is 'theories in action' rather than espoused theories that count.

Roth et al. (2007) describe their experience of acting in the political landscape and come up with four strategies:

1 Anchor the project in the organization and find the right stakeholders and sponsors.
2 Be 'street smart' and know how to get things done.
3 Ensure that the first interventions and change in the project have a wide impact on a wide range of organizational members.
4 Describe the action research projects as organizational change programmes.

Managing political relationships

In order to be able to manage the content and control agendas of the action research project and the power-political processes of influencing and ensuring the legitimacy of your project, you need to be able to manage your superiors, peers and colleagues. We have identified 10 key power relationships. All of the 10 key power relationships need to be considered and managed when carrying out a quadrant 3 type action research project where the system is not committed to self-study in action (see Chapter 8). The first two relationships may not be of great importance for quadrants 2 and 4 research projects where the system is committed to self-study in action.

1 *Your relationship with your sponsor.* It is most likely that you, if you have a middle or low organizational rank, have a sponsor, one who provides permission and primary access to undertake the research, both in the initial and in the latter stages of the project. Where the research is part of a degree programme, the sponsor grants you permission to have time off to attend the course, take study leave and use organizational materials for research. The sponsor may be your immediate superior within the same department. In this case the relationship may be close and supportive. The sponsor may be elsewhere in the organization, in a position of higher management. You need to work at maintaining this relationship as the continuation of the research project may depend on it. This may become particularly difficult if it emerges that the sponsor is a source of problems within the organization. You need to keep your sponsor abreast of developments and seek his or her counsel. That way you keep him or her informed and on your side.

2 *The sponsor's relationship to other executives.* Your credibility and access may depend on the sponsor's status and standing within the organization. If the sponsor is not considered favourably by other executives, this may have a negative effect on how your research project is perceived. The sponsor's power relationship with other powers in the organization is critical in gaining acceptance for that research from higher levels of management or administration. Secondary access may be granted or denied at this level. You may have to leave the sponsor to do your access negotiation for you or you may be allowed to approach these other executives yourself. This will depend on the nature of the project. It may be that you are helpless in this regard. Whatever the project, you will have to work independently at establishing the credentials and value of the research project.

3 *The relationship of executives with each other.* The power dynamics between departments or individual heads of departments may be a relevant feature in promoting or blocking the research. If you are from one department, that may inhibit cooperation from other departments. This may be the most significant political force for you as an insider researcher, and the one over which it may be most difficult to exercise control. The key is to build personal relationships with significant persons in other departments so that they will cooperate. Perhaps some of them will be members of your project team.

4 *The relationship between you and significant others.* Whatever the relationship between the sponsor and other executives, you must be able to establish your own relationship with significant others, many of whom may be key executives. This is particularly relevant if you wish to interview senior executives and ask what might be experienced as awkward questions. If your sponsor falls from favour, you will need to have established relationships with significant others in order to maintain your profile and project.

5 *The relationship between executives and others in higher management.* Senior management at a corporate level may undermine the research or withdraw consent. The relationship tends be to be remote, in that the executives are not likely to know you personally. It is usually unnecessary for them to have any detailed knowledge of your project. You may not have access to these people, so you may find it difficult to exercise influence over them.

6 *The relationship between executives and organizational members.* This relationship includes relationships between management and workers, and management and trade unions, so that the research is accepted by the relevant parts of the organization. The research project may fall victim to ongoing organizational relationships, where employees use a lack of cooperation with you as a power tool to express dissatisfaction with some unrelated aspect of organizational life over which there is dispute in order to gain political leverage. In these situations you are powerless and dependent on others for the resolution of the dispute.

7 *The interdepartmental relationships.* Some departments have more power than others, and there are different subcultures, all of which may work for or against a research project. If, for example, you work in the head office or corporate centre, you may have to deal with the attitudes of those in regional offices who view anyone from the corporate office with a prejudiced eye. The key is to establish a personal relationship with significant persons who will cooperate with you.

8 *The relationship between the researcher and subordinates.* You may be relying on subordinates for significant information. They may feel the need to be less than honest with their boss, who is undertaking the research. It may be that your own behaviour and management style is a critical factor in the issues under investigation, and so subordinates may be reticent in providing accurate information or feedback. In such a case, where you are a superior, having a third party gather data may be essential.

9 *The relationship with customers or clients.* These people may be the ultimate beneficiaries of the research or may be involved in the actual research process. Approaching clients and customers has political complexities as it may raise expectations about the service provided to them.

10 *The relationship between you and your peers.* Engaging in research which involves your peers, some of whom may be friends, is particularly sensitive and may make the research process stressful. If peers and colleagues are the subject of observation and comment, they need to be informed and actively involved. They need to be protected from possible retaliation by superiors. At the same time, you need to be wary of how you may be biased in favour of peers and colleagues, be seduced by the closeness of the relationship and hence be unable to reflect and critique. Peers, colleagues and superiors may be asking the following sorts of questions: what are you observing? What are you writing about me? Am I being criticized? For those unfamiliar with an action research approach, the idea of doing research in the everyday job may be hard to grasp. How can you work and include research at the same time? This question comes not only from a limited notion of what constitutes research, but also from a fear of being criticized in writing behind one's back.

The management of an insider action research project involves attention to significant relationships by building support and involving key others. If the CEO wanted to be a member of your inquiry group, would it work? Would you be able to tell him that you think it is a bad idea?

Bjorkman and Sundgren (2005) reflect on being political entrepreneurs and relate their respective insider action research projects to the 10 key relationships. They note that finding red-hot issues is closely linked to relations with sponsors and executives, to the sponsors' relationship to other executives and to the relationship of executives with each other. They refer to building a *relational platform* to address the needs of managing these relationships directly. They suggest extending the list of 10 power relationships further to include the utilization of external networks.

Friedman (2001) reflects that in cases of individual research where issues are non-technical, there are likely to be insurmountable obstacles and high degrees of defensive routines. How then do you work as a 'political entrepreneur' engaging in the public performance and backstaging activities? Kakabadse (1991) presents six useful guidelines:

1 *Identify the stakeholders.* This means identifying those who have a stake or interest in the project and its outcomes and approaching them so as to identify their intentions.
2 *Work on the comfort zones.* This means working on those behaviours, values and ideas which a person can accept, tolerate or manage. As long as these are not threatened, people will be able to focus on wider concerns.
3 *Network.* This means going beyond formal hierarchies or structures where necessary to coalitions of interests which may exert greater influence on key stakeholders than the hierarchical structure.
4 *Make deals.* Making deals is common in organizations as individuals and groups agree to support one another on a particular issue in return for support on others. This is a common way of reaching agreement on policies.
5 *Withhold and withdraw.* It may be useful on occasion to withhold information in order to not to fuel opposition, though you would not want to withhold information constantly. It is also useful on occasion to withdraw from conflictual situations and let others sort out the issue.
6 *If all else fails.* Kakabadse recommends that you need to have some fallback strategies if all else fails. These obviously depend on the demands of the situation and what you can personally handle and manage.

Friedman (2001) provides more specific guidelines:

1 Describe your own reality image and situation as concretely as possible.
2 Ask senior and middle management if this explanation accurately fits as they see it.
3 If there are significant differences, inquire into the sources of these differences.
4 Continuously inquire into the reasoning behind actions.
5 Design strategies dealing with the current situation and similar future ones.

Ramirez and Bartunek (1989) reflect on role conflict explicitly in their case of an insider action research project in a healthcare organization. They cited two specific instances. In one instance, they noted that the insider action researcher had to deal with the twin role of facilitating meetings while at the same time acting as a department head whose status was junior relative to other participants. The second role conflict was more explicitly political. Other organizational members spread rumours about the action researcher to the effect that she was engaging in the research to set up a position for herself. The researcher's experiences of being the recipient of such political behaviours caught her off guard and were hurtful to her.

Researchers often think that they have little power in the research process because they are dependent on powerful others for access. Others may see the researcher as powerful because he or she is knowledgeable, has initiated the research and is selecting whom to involve. In effect they may see the researcher's view of reality as being given public visibility. Accordingly, therefore, we need to examine some ethical issues of doing action research in your own organization.

Williander and Styhre (2006) implicitly explore the nature of the social construction of Williander's project through reflecting and how it was perceived by senior managers. The lack of a budget could be said to be a reason why the project could not start but the lack of a budget could be due to managers' unwillingness to budget. As long as the insider action researcher had difficulties in expressing the specific need and resulting transformational state of the suggested research project sufficiently explicitly, and preferably in financial terms, it remained too vague for senior management to act on it. They report that one manager referred to the project as 'your scientific hocus-pocus', while another suggested that 'This sounds too theoretical and time-consuming – why don't we just come up with some innovations?'

Questions for reflection and discussion

- Does this reaction to Williander's action research initiative sound familiar? What is your experience?
- How might you seek to address and change the constructions that others put on your project?

Ethics

Ethics procedures are part of life and so they are part of research, and increasing strictures are being placed on research by universities and organizations through institutional review boards or ethics committees. In some settings researchers are required to fill out an ethical approval form prior to commencing the work. Over the past few years a literature on ethics in action research has emerged that may help you to think out the issues you face and to design your research (Coghlan and Shani, 2005; Brydon-Miller, 2008). The special issue of *Action Research* devoted to the subject of ethics and action research is a particularly valuable source (Brydon-Miller et al., 2006), as are the other works referenced in this chapter. Since action research has an unfolding nature as it attempts to integrate inquiry with everyday organizational action, one may argue that the ethical issues of action research are not different from the ethical issues of a good life. Eikeland (2006b) distinguishes between philosophical and applied levels of ethical aspects of action research. Action research is grounded in principles of democracy, justice, freedom and participation, though as Boser (2006: 14) comments, 'Democratic intentions do not obviate the need for thoughtful examination of the ethical implications of the research on individuals and other stakeholders groups.' Hilsen (2006) argues that ethics in action research may be based on three pivots: human interdependency, cogeneration of knowledge and fairer power relations.

Quality of relationships

Rowan (2000) presents a series of five concentric circles that describe ethical issues in different forms of research in psychology. The first circle is the 'natural inquiry' arena in which researchers engage in what we typically term positivist research. Here the researcher and participant engage at a single of point of contact, namely the researcher's agenda to obtain information from the participant. The ethical issues which arise in this one sided relationship typically focus on doing good, not doing harm and respecting the person (what Eikeland, 2006b refers to as 'condescending ethics'), and these issues are met through communication. Participant observation, which often involves disguising the purpose of the researcher's involvement, does mean that deceit is part of the research process. The second circle is the 'human inquiry', which is captured by the hermeneutics and phenomenology. In this approach the participant is treated as one who is fully human and not simply as a set of variables to be measured. There is emphasis on empathy, identification, trust and non-exploitative relationships.

For Rowan there is a major shift when one moves to the third circle. Rather than the researcher meeting participants at one carefully stage-managed point of the research process, the researcher involves participants in planning the research and in processing the results. This is the arena of action research. Ethical issues concern not only individuals, but also the whole community or organization. Unintended consequences are a deliberate focus. The researcher's self-understanding and social vision come into play, with issues of power central to the process. The fourth circle is transcendent research which focuses on transpersonal research, spirituality and mindfulness (Bentz and Shapiro, 1998). Ethical issues attend to the spiritual level in persons. Rowan's fifth circle is comprehensive, systematic inquiry which encompasses all the others and which demands being able to think and act appropriately in each research setting, with a concern for the researcher, the participants and the wider system, taking into account the spiritual implications of what is being undertaken. Rowan provides a framework for thinking about the forms and quality of relationship that action researchers have with participants in different research settings. It directs the ethical behaviour in which action researchers must engage in the democratic, participative values on which action research is grounded.

Williamson and Prosser (2002) pose three ethical questions which, in their view, action researchers and participants need to be clear about, discuss and agree the answers:

1 If researchers and participants collaborate closely, how can confidentiality and anonymity be preserved? As action research is a political enterprise and has consequences for participants and the researchers, it is difficult to guarantee anonymity and confidentiality as others can easily know who participated and may be able to identify who said or contributed what.

2 If action research is a 'journey' and 'evolves', how can informed consent be meaningful? Neither action researcher nor participants can know in advance where the journey will take them and cannot know to what they are consenting. As a change process can create its own resistance, action researchers cannot be expected to withdraw in the face of opposition (albeit by small groups within the project).

3 As action research can have political consequences, how can action researchers avoid doing harm to participants? Williamson and Prosser point to two ways of answering this question: the establishment of an ethical code for action researchers, and the extent to which the collaboration and negotiation occur so that participants own the findings as much as the researcher.

Walker and Haslett (2002) ground the issues of ethics in action research in the action research cycle itself. They suggest that ethical questions may be posed in terms of possible and actual ethical questions around the cyclical activities of planning, action and reflection. Processes of obtaining consent, ensuring anonymity and confidentiality, and balancing conflicting and different needs, are actualized in planning, taking action, collecting data and interpreting. They cite Stringer's (2007) two important questions as central questions running through the whole project: who will be affected? How will they be affected? Gellerman et al. (1990) articulate four ethical principles:

1 Serve the good of the whole.
2 Treat others as we would like them to treat us.
3 Always treat people as ends, never only as means; respect their being and never use them for their ability to do; treat people as persons and never as subjects.
4 Act so we do not increase power by more powerful stakeholders over less powerful.

Protocols and ethics committees

In general, the role of ethics committees and institutional review boards is to avoid or prevent abusive behaviour and to prevent violations and lawsuits against universities, and so they act as guardians of ethical practice. By and large, the members of these committees come out of the conventional positivist research tradition and expect that hypotheses, methods and expected outcomes will be well articulated in advance so that review of research proposals is fairly straightforward. When they are confronted with action research proposals they are frequently at a loss how to understand this form of research and how to evaluate a proposal (DeTardo-Bora, 2004). Given that action research is an unfolding, emergent process which evolves through cycles of action and reflection, it is not feasible to map out a detailed anticipation of ethical issues in advance which will cover all eventualities (Morton, 1999; Walker and Haslett, 2002). Lincoln (2001) suggests that protocols are inadequate and are insufficient to meet the face-to-face, participative close work of action research. At the same time, it is possible to articulate some ethical principles to guide your work as an action researcher. Boser (2006) proposes that attention to ethics needs to:

1 be guided by a set of externally developed guidelines that direct attention to the set of relations among those participating in or affected by the research
2 be integrated into each stage of the action research cycle to inform decision making by stakeholders
3 be transparent to the larger community.

Brydon-Miller and Greenwood (2006) are optimistic about the difficulties in going before such committees and suggest that action research actually holds out more guarantees for the ethical treatment of human subjects than conventional research does. This is because action research is built on a voluntary partnership with stakeholders who form a collaborative team, learn and apply the methods together, implement the methods together and analyse the outcomes together. They suggest a dialogue between the faculty that is supporting the action research and the ethics committees and institutional review boards. This dialogue may fruitfully explore the dialectic tensions between predictability and process, protection and participation, confidentiality and credit, coercion and caring.

Insider action research and ethics

As you are an existing member of your organization and probably have multiple roles to which you are adding the researcher role, as we explored in the previous chapter, role clarification around your researcher tasks, such as gathering information and analysing data, is complex (Holian and Brooks, 2004). In addition, not only may the notion of action research be strange for ethics committees and institutional review boards, but the idea of doing action research in your own organization may be even stranger. Joan, an insider action research student, when her research proposal was being turned down by the ethics committee for the second time, retorted to the committee: 'You are telling me now that I cannot do my job, the job that I have been doing for eight years. What I am seeking approval for is what I do every day.' Joan's dilemma with regard to doing action research within her own organizational role clarified a central point with regard to ethics approval for insider action research, a point reinforced by Brydon-Miller and Greenwood (2006):

- A distinction needs to be made between engaging in action research and reporting on that action research. The action research activity may be integral to your day-to-day work and as such does not require ethical approval. What does require approval is the process of taking this action and transforming it into research for publication or academic credit. This distinction is crucial and it is for the latter that ethical approval may be directed.
- Faculties needs to work with the ethics approvals committees to incorporate the development of proposal forms and consent forms into the action research process and for the forms to incorporate the action research paradigm. This is a process of collaboration in action, and action research faculties are on familiar ground in engaging on this issue.
- Ethical approval is only one step in the process and, as Boser (2006) argues, ethics needs to be integrated into each stage of the action research cycle. So the questions and issues identified above are critical to keep in mind as the research progresses.

Humphrey reflects:

> I tottered on the tightrope of the insider-outsider hyphen, torn between the views and values of academics and activists camped on either side of me and overlooking the murky waters below. There is no Code of Ethics which could have offered a safety net in this situation and as I found myself sacrificing some principles to others, I consoled myself by repeating the mantra it was only my clothing that had become mud-splattered rather than my core self. (2007: 16–17)

Integrating roles, politics and ethics

There are ethical dilemmas attached to how action researchers hold their researcher and their organizational action roles. In terms of the organizational action, they are bound to provide a quality service to the organization's management; as researchers they have a responsibility to go beyond the boundaries of the particular project to contribute to the generation of knowledge. In addition to the ethical dilemmas pertaining to these separate roles, there are dilemmas which pertain to the integration of the roles which are held together by action researchers (Benne, 1959; Lippitt, 1961; Kelman, 1965). Applying Katz and Kahn (1978), White and Wooten (1986) refer to a 'role episode', by which they mean studying an ethical dilemma through an ambiguity or a conflict between the sending role and the receiving role. The 'sending role' comprises expectations of what the action research role will fulfil. The 'receiving role' has perceptions of both its role and the sending role and then either complies or resists. As White and Wooten (1986) illustrate, this role episode model is a useful one to investigate and explain the behaviour of change agents and client systems, and as such is relevant to our consideration of action researchers. They apply the role episode model to five categories of ethical dilemma: misrepresentation and collusion, misuse of data, manipulation and coercion, values and goals conflict, and technical ineptness.

With regard to the dual role of action research, Morton (1999) describes four ethical dilemmas in terms of 'role contamination':

- What might action researchers promise clients? Action researchers should not promise beyond what they can reasonably deliver, yet the process of action research is innovative and involves a degree of risk. So action researchers have a dilemma in holding both promise and risk.
- How theoretical can action researchers be on organizations' time and money? Action researchers, more than consultants, value reflection and theorizing; these satisfy their own intellectual interests but may be of little value to organizational management. Action researchers need to balance their research-oriented activities with their action-oriented activities.
- How do action researchers present themselves? Presenting themselves primarily as academic researchers who also do consulting, or primarily as consultants who have academic interests, each has its own pitfalls, depending on the audience and the audience's view and expectation of either role.
- How can action researchers resolve the conflict between the quality of the organizational action and the quality of research? While one expects them to be in harmony, this may not be necessarily so. A failure in the organizational action may generate rich research data. If an organizational action is heading for possible failure, action researchers may be confronted with having to put priority on one over the other.

Humphrey concludes her reflection:

> It is the activation of the hyphen which enhances one's chances of surviving and thriving in complex territories. The perils of the inside-outsider are that she can be pushed and pulled along an invisible insider-outsider continuum by others who have a vested interest in who she is and what she is doing, and she is cast now as an 'insider' and now as an 'outsider' by different actors and audiences and can lose her sense of herself. To actively take charge of the hyphen is to appreciate one's uniqueness as an insider-outsider and to cultivate the art of crossing between life-worlds. It is only when the researcher cherishes herself as an insider-outsider and commits herself to journeying between life-worlds that she can protect herself and her project from others. (2007: 23)

Journaling is a most valuable tool for coping with and exploring political and ethical issues. As described in Chapter 3, it can be a vehicle for dumping painful experiences as well as for articulating them and reflecting on them in a more reflective space.

Humphrey describes what her journaling meant for her:

> It is not surprising that my journal came to function as a 'friend' in the sense of 'confidante', since the more I censored and silenced myself with my comrades, the more I disclosed to my journal. A journal absorbs our disturbing thoughts and feelings without pronouncing judgement upon them; it allows us to draw pictures in accordance with the language of our body and to write poetry in accordance with the language of our soul; it also stores these archives for our own subsequent processing. My journal contains references to being 'shipwrecked on a desert island' and 'seeking refuge in the desert'. (2007: 18–19)

Integrative case

Holian (1999) provides a case which integrates some of the themes of Chapters 9 and 10. She reports how her additional researcher role added a complex dimension to her senior executive role. She found that when organizational members provided information to her 'in confidence', there was some doubt as to whether it was in confidence to her as a researcher or as a senior executive. Merely asking informants as to which hat they saw her wearing at the time did not resolve the uncertainty. If information was provided to her as a senior executive, she might have been authorized or even obliged to act on it to prevent harm to others. If it were provided to her as a researcher, she might not have the right to do so. Whatever the role, organizational members knew she was the same person and knew what they had told her and that she could not forget it.

The role conflict between her senior executive position and her action researcher role that she experienced when organizational members provided her with information which she did not know if she could use in her researcher role, and which she thought she should use in her executive role, created an ethical dilemma for her. As her research subject was ethical decision making she faced a double dilemma: a content one for her organization and a process one for the research. She established and participated in a cooperative inquiry group comprising people in decision making roles from a diverse range of organizations. The members of this group discussed ethical issues they were experiencing and encouraged one another to reflect on their own experience and find new ways of working with ethical issues in their own organizations.

She reported how she felt unprepared for the backlash which resulted from surfacing 'undiscussables' within the organization related to cover-ups, perceived abuses of power, nepotism, harassment, allocation of rewards ands unfair discrimination. While these issues were deeper, more shocking and more troubling than anticipated, she reflected that she was not adequately prepared to look after herself or others when the backlash came. Consequently, she was not able to balance the multiple roles of researcher, senior executive and programme facilitator and, after one last stand-up fight with some of her fellow senior executives, she resigned.

Authenticity involves attention to and reflection on the personal questions and dilemmas which arise in the political dynamics of the action research projects. The outer dynamics pertain to attention to the political behaviour of others, while the inner dynamics refer to how you think and feel about engaging in political activity. Being intelligent refers to understanding the structural patterns of political action and how they are played out in this organizational setting. Being reasonable means weighing the evidence and judging the demands of political action. Being responsible means you take responsibility for what you decide to do and what you do in second person action.

Conclusions

When you are engaging in action research in your own organization, politics are powerful forces. You need to consider the impact of the process of inquiry, who the major players are, and how you can engage them in the process. Ethics involve not merely not deceiving or doing harm, but being true to the process. This does not mean telling everyone everything or being politically naive, but rather it means recognizing who the key political players are and how they can value the research by participating in it. It may seem that political dynamics are the major obstacle to doing action research in your own organization and it may put you off. At the same time there are those who revel in political behaviour and enjoy the cut and thrust of attempting to make a difference through their action research project.

Exercise 10.1 Assessing politics and ethics

Politics and ethics

Political issues

1 Challenging the status quo: how can you do things differently?
2 Changing existing power relations: do you have the power to change this for yourselves?
3 The system reasserting existing power relations: you don't have the authority to change this for yourself.

Ethical dimensions

1 When you collaborate with others, how can confidentiality be maintained?
2 If your action research project is a 'journey' and 'evolves', how can informed consent be meaningful?
3 As action research can have political consequences, how can you avoid doing harm to others?

Levels of inquiry

As a concrete application of Table 10.1 to your project, answer the following questions and fill out the table for yourself.

	Task	Process
First person	What are your specific political and ethical tasks in your project?	How are you to achieve your goals ethically and politically?
		(Continued)

(Continued)

	Task	**Process**
Second person	Who are the political stakeholders in your action research project? What interests do they have that need to be served?	How do you work with the different stakeholders to achieve the purpose of the project?
Third person	What knowledge are you generating from this project that informs your understanding of the political dynamics of organizations?	How do you build this actionable knowledge about the political dynamics of organizations from your first and second person inquiry and practice?

Exercise 10.2 Force field analysis

Lewin's force field analysis is a most useful tool for assessing and constructing interventions with respect to organizational political forces. As it is presented in many organization development texts, force field analysis is a technique created by Kurt Lewin for problem solving or managing change. It is based on the assumptions that in every situation there are forces driving change and forces restraining change, and that an emphasis on reducing restraining forces is more effective than increasing driving forces. While a force field might look like what we might do in listing reasons for and against taking an action, it is actually quite different. Reasons for and against are static and rational; they have to be justified. In force field analysis, forces impinging on a situation are listed. Hence, with regard to organizational politics a force field of political driving and restraining forces may provide you with a useful insight into what is going on and help you construct interventions to reduce restraining forces.

Force field analysis comprises five steps:

- *Step 1*: describe the change issue and the desired direction of the change.
- *Step 2*: list the political forces driving change and those restraining change in a diagram which has the forces in opposition to one another.
- *Step 3*: give a weighting to the forces, identifying those that are stronger and more powerful than others.
- *Step 4*: focus on the restraining forces and assess which of the significant ones *need* to be worked on and which *can* be worked on.
- *Step 5*: develop plans for reducing these forces.

ELEVEN

Writing Your Insider Action
Research Dissertation

In this chapter we explore action research in the context of engaging in it for academic accreditation, such as a master's or doctorate. In recent years there have been valuable publications on action research dissertation writing, notably Dick (1999), Zuber-Skerritt and Perry (2002), Herr and Anderson (2005), Coghlan and Pedler (2006), and Zuber-Skerritt and Fletcher (2007). We draw on their work and bring our own reflection on insider action research to bear on their respective work.

Remember that action research is a different form of research than traditional research:

- traditional research begins with what we know and seeks to find what we don't know
- action research begins with what we don't know and seeks to find what we don't know
- what we don't know that we don't know is the particular fruit of action research.

As we discussed in Chapter 2, there are two action research projects running concurrently, what Zuber-Skerritt and Perry (2002) call the *core* action research project and the *thesis* action research project. Now we are focusing on writing up the *thesis* action research, which is the inquiry in action into how the *core* action research project was designed, implemented and evaluated and how you enacted your role in it and how you now reflect on it.

An action research dissertation

At the end of the academic-oriented action research project you have to write a dissertation. In non-academic contexts you may write a report or want to write an article or paper. We assume that you have been writing all through your project, as you have written up accounts of events and kept your own personal journal and reflections up to date. It is important that you keep records and notes in real time, close to events, so that you have an accurate record of

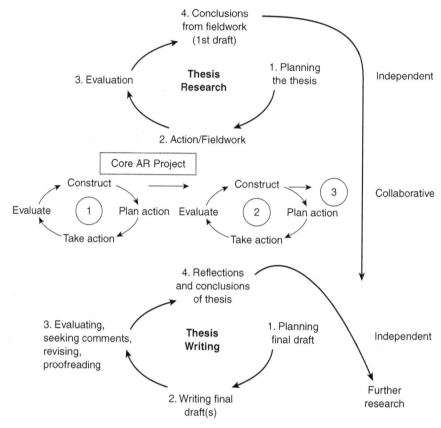

Figure 11.1 Conceptual model of an action research thesis (adapted from Zuber-Skerritt and Fletcher, 2007: 421)

what took place, what you were thinking about them at the time, and the insights you had and the judgements you made.

A dissertation is an academic document and therefore needs to conform to academic requirements around justification of topic and approach, description and defence of rigour in methodology and methods of inquiry, familiarity with existing content and process literature, and contribution to knowledge. An action research dissertation is no different, though its presentation and argument differ from traditional presentations.

Zuber-Skerritt and Fletcher (2007) elaborate the interrelationship between the *core* action research and the *thesis* action research. From Figure 11.1, you can see that the *core* action research project is a collaborative venture whereby the cycles of action and reflection are enacted in multiple successive and concurrent cycles in first and second person practice, as we discussed in Chapters 2 and 3. The *thesis* research project involves you in independent work, at both ends of the entire process. In this chapter we focus on the thesis writing cycle.

Constructing and writing your dissertation

Practices describing action research dissertations typically suggest that they be structured to deal with:

- purpose and rationale of the research
- context
- methodology and methods of inquiry
- story and outcomes
- discussing quality
- self-reflection and learning of the action researcher
- reflection on the story in the light of the experience and the theory
- extrapolation to a broader context and articulation of usable knowledge.

This is not to say that such a structure would necessarily mean that each of these headings has to be a chapter in itself; rather, these issues need to be clearly dealt with formally. For example, the story may be spread over several chapters, depending on its length and complexity and the extent of the research project.

Purpose and rationale of the research

When you present the purpose and rationale of your research you are, in effect, presenting its academic context. This involves stating why the action you have chosen is worth doing, why it is worth studying and what it is that it contributes to the world of theory and practice. The most critical issue for you at the outset of writing an action research dissertation is to make an academic case for what you are doing. This is not just an argument for credibility but a formal effort to locate your work in an academic context.

Context

We have outlined the main strands of a discussion of context in Chapter 4. While then you were outlining it in a proposal, here you are writing about it in depth. Academic context is also important. Not only are you reviewing the social and organizational context of your project but you need also to review and critique the research that has been done in that context. We refer to this literature as *outset* literature, that is, the literature that you read, discuss and critique at the outset of your research and which helps ground your research in existing work and opens the door to your contribution.

Williander and Styhre (2006) outline the main contextual issues that underpin Williander's insider action research work in creating the eco-benign car in Volvo. They explicitly discuss external factors pertaining to the car manufacturing industry and to the challenges facing Volvo. They conclude that

Volvo needed two types of change: what they refer to as 'locking-up the minds from the lock-in effects of the technological paradigm in order to be more objectively able to assess the possibilities of going green' and 'the creation of the capabilities required to exploit these possibilities' (2006: 243).

Methodology and methods of inquiry

This is your major section on methodology in which you outline and justify your approach. Here you describe your action research approach, methodology and methods of inquiry. Methodology is your philosophical approach; methods describe what you actually did. Accordingly, you need to articulate both your methodology and your methods of inquiry.

As with any research dissertation, you need to review the theory and practice of your methodology, in this instance action research. This is a matter of providing definitions, some history and its main philosophical tenets. Second, you also then need to review the practice of action research in your own field, such as in nursing, education, information systems research, and so on. Third, you need to describe and review the particular approach within action research that you might be using, particularly if you are using one approach predominantly. Accordingly, for example, you will review and critique the theory and practice of appreciative inquiry, cooperative inquiry, action learning, and so on, as appropriate.

Regarding methodology, you need to convey that you are using a normal and natural research paradigm with a long tradition and adequate rigour which is suitable for the project on which you have worked. As Dick (1999) very usefully points out, it is important to argue positively for your approach, rather than to criticize negatively the limitations of other approaches. As we presented in earlier chapters, the action research literature provides extensive justification of how action research is scientific and rigorous.

While all research demands rigour, action research has to demonstrate its rigour more particularly. This is because in action research you typically start out with a fuzzy question, are fuzzy about your methodology in the initial stages and have fuzzy answers in the early stages. As the project develops, your methods and answers become less fuzzy and so your questions become less fuzzy. This progression from fuzziness to clarity is the essence of the spirals of action research cycles (Gummesson, 2000). Accordingly, you need to demonstrate clearly the procedures you have used to achieve rigour and defend them. As Dick (1999) emphasizes, this means that you show:

- your use of action research learning cycles
- how you accessed multiple data sources to provide contradictory and confirming interpretations

- evidence of how you have challenged and tested your own assumptions and interpretations continuously throughout the project
- how your interpretations and outcomes are challenged, supported or disconfirmed from existing literature.

Discussing quality

As we discussed in Chapter 1, it is important to be explicit about how you have tried to ensure quality in your action research project. There several frameworks which are useful in establishing quality criteria and for exploring quality in action research and which can be applied to your dissertation work (Levin, 2003; Herr and Anderson, 2005; Reason, 2006; Zuber-Skerritt and Fletcher, 2007).

In terms of Shani and Pasmore's (1985) complete theory of action research, the quality of your project rests on how you have presented:

- context
- quality of relationships
- quality of the action research process itself
- outcomes.

This is a simple four point framework that is easy to use, particularly in a master's thesis. An equally simple framework is Levin's (2003) four criteria: participation, real-life problems, joint meaning construction and workable outcomes. With regard to participation, how well does the action research reflect cooperation between you and the members of the system? With regard to real-life problems, is the action research project guided by a concern for real-life practical outcomes and is it governed by constant and iterative reflection as part of the process? With regard to joint meaning construction, is the process of interpreting events, articulating meaning and generating understanding a collaborative process between you and the members of the system? Finally, with regard to workable solutions, does the action research project result in significant work and sustainable outcomes?

Herr and Anderson (2005) use the term 'validity' and explore five types: outcomes, process, democratic, catalytic and dialogic. In terms of Reason's (2006) choice points, you may ask how your study may be judged in terms of:

- being explicitly both aimed at and grounded in the world of practice
- being explicitly and actively participative, that is research *with*, *for* and *by* people rather than *on* people
- drawing on a wide range of ways of knowing – including intuitive, experiential, presentational as well as conceptual – and linking these appropriately to form theory
- being worthy of the term 'significant'
- emerging towards a new and enduring infrastructure.

As Reason (2006) points out, your research cannot rate equally highly at all of these, so he argues that as an action researcher you need to be aware that these are choice points. Accordingly, you can select which ones you want to be judged against and make them clear and transparent.

In your section on methods you describe how you are doing action research. Methods of inquiry refer to the content and process issues of how you framed and selected the issue; how you built participation and support; how you accessed and generated data; how you recorded data, such as notes and minutes of meetings, journal, where and how you used observation, interviews or survey instruments; how you engaged others in the action research cycles of implementing the project; how you dealt with the political and ethical dimensions presented in Chapter 10, particularly with regard to reflective learning and so on. These are all pertinent issues.

Story and outcomes

The heart of the dissertation is the story or course of events. At the initial stages, you are likely to construct the story around a chronological narrative and structure it in terms of significant periods or particular projects. So at the draft stages you might have narratives which cover particular periods or particular projects. This is an important structure to follow at the early stages of your write-up as it enables you to get the story down on paper in a logical sequence. The next stage of writing the story is to reflect on it and test your insights as to what themes emerge. Then you may find that you are surfacing images or themes for the periods or projects which capture your sense of the meaning of the project and lead you to a synthesis.

Action researchers are often surprised at what happens during the writing of a dissertation. They think that it is simply a mechanical task of writing up what is in their notes and files. Experience shows that the writing up period is a whole new learning experience. It is where synthesis and integration take place. From what hitherto have been isolated masses of details of meetings, events and organizational data, notes on scraps of paper and disks, notes from books and articles, a new reality emerges. Things begin to make sense; meanings form. For many researchers, this is the time they realize what they have been doing all along.

Writing the story is the key to synthesis. You are likely to have far more detail than you need or can use. Therefore, as you begin to select what to include and exclude, you are beginning to form a view of what is important in the story. You are at the next stage of reflective practice, and indeed of action research itself. The writing project becomes an action research project as you engage in cycles of drafting and revising, seeking comments from your supervisor, reflecting, understanding how what you have written fits into the whole, and then formulating conclusions (Zuber-Skerritt and Perry, 2002; Zuber-Skerritt and Fletcher, 2007). It is far from being the mechanical task of writing up your notes that you might have thought.

A critical issue in presenting the story is to distinguish the events which took place, about which there is no dispute, and the meanings you attribute to these events. It is important to present the basic story separately, as if it were a news bulletin or as if a video camera had recorded what had taken place. As McTaggart (1998) points out, the narrative needs to be sufficiently comprehensive and transparent so that the reader can arrive at the end of it able to judge for him/ herself the validity of the research, its claims to the creation of knowledge and any claims for its transportability. This form of presentation gives the evidence in a factual and neutral manner. Your view of these events and your theorizing as to what they mean should not be mixed in with the telling of the story. This should come later, perhaps at the end of the chapter or the end of a particular phase of the story. By separating the story from its sense making, and by clearly stating which is story and which is sense making, you are demonstrating how you are applying methodological rigour to your approach. Combining narrative and sense making leaves you open to the charge of biased storytelling and makes it difficult for readers to evaluate your work.

Winter (1989) argues that your presentation should reflect your own process of learning and not be a judgement of others. He makes the relevant point that you should avoid making commentaries which place you as the researcher in the superior role of one whose analysis of other people's words shows that you understand what took place while they don't. He suggests that your commentary should place you at risk, as the single voice of the author in a context where many people participated in the work.

> A useful technique for using your own reflections as you are telling the story is to insert *a reflective pause* in a box at periodic intervals, giving your own reflections on what you have just recounted. Here you can reflect in public, show your train of thought and your insights and share what you plan to do next and why. This technique provides a mechanism for having accounts of your own reactions, interpretations and action planning alongside factual narrative in a way that doesn't confuse the two. The reader then can see what is happening in the story, can read what you are thinking and how you are interpreting the situation as it unfolds, and can follow the logic of your thoughts and actions. This is a tool for inserting first person narrative alongside the second person narrative. It provides a concrete demonstration of your use of general empirical method and your quest for first person authenticity.

In an action research dissertation, your account of your sense making often takes three forms. One is where you make sense of particular events within the narrative as it progresses. The *reflective pause* technique described above is one such form. Bourner (2003) suggests that to demonstrate reflective learning and critical thinking it is useful to note how events 'surprised' you, 'disappointed' you and so on. These notes, which may be shared in the reflective pauses, illustrate the application of your first person inner arc of attention to second person external events and to the processes that create third person insights. The second form is at the end of a chapter or section where you may present how you make sense of that event. You do that clearly so that the reader may know what you are doing and go back to the story to see how your interpretations make sense. It is important that your sense making sections are not too far from the story narrative. If you leave all your sense making to chapters at the end, particularly in a doctoral dissertation, it makes it difficult for the reader to remember to what you are referring. The third form that sense making takes is that you have a general chapter towards the end of the dissertation which integrates the more specific interpretations you have made, provides an overview of your sense making of the whole story, and links it back to the purpose of the research, the context and the contribution to actionable knowledge.

It is at this stage that you may have to draw on a further content literature. This is what we call *emergent* literature. As you are progressing through the story and making sense of the story you will find that you are being drawn into more specific or even new areas of content and process, whose literature you now need to read and review. In action research projects, specific relevant content areas emerge as the project progresses, so you often don't quite know what the focus of your synthesis is until the project is well in progress. Content literature becomes more focused through the story and directly relates to what is being framed in the story. Angela, whom we encountered in Chapter 4, found that as her action research progressed the team dynamics became central. So she read some books and articles on teamwork and stages of team development to help her make sense of what was going on in her team and to assist her in her team role in leading the project. This was an area of literature that emerged in her project and could not have been anticipated at the outset.

Self-reflection and learning of the action researcher

An important part of the action research dissertation is your reflection on your own learning. As you have been intervening in the organizational system over the period of the action research project, you need to articulate what you have learned, not only about the system you have been working to change, but about yourself as an action researcher. The project may have challenged many of your assumptions, attitudes, skills, and existing organizational relationships, as we have seen in the cases of Humphrey (2007) and Moore (2007). Many of these points of learning have been expressed in the reflective pause boxes, discussed above, which try to capture your reflection in action. This first person material is important as it contributes to the integration of the three voices – first, second and third person.

Reflection on the story in the light
of the experience and the theory

One of the most common criticisms of published action research is that it lacks theory. In other words, action research accounts tell a story, but don't address issues of emergent theory and so contribute to knowledge. Accordingly, your action research project needs to apply some established theory or extend or develop theory. The scope of the academic project, whether master's or doctorate, is an important determinant of what is to be expected in this regard.

If you are a participant in a master's programme engaging in classical action research, such as an action-oriented MBA programme, you use frameworks to make sense of what is going on. You may be drawing on frameworks which help you make sense of an industry analysis, the performance of a firm and the like. Your use of these frameworks aligns the story to the theory, and through this alignment you demonstrate your understanding of the theory and its application.

If you are engaged in a more research-oriented programme, such as a master's by research or a doctorate, you are not only aligning the story with theory, but also extending that theory or developing it. This extension is an inductive process, coming out of your meta learning of reflecting on the implementation of the action research cycles with the members of the systems as they enact the action research project. This extension or development of existing theory may be in content or process.

Extrapolation to a broader context and
articulation of usable knowledge

As a consequence of your reflection on the story and articulation of usable knowledge, you need to articulate how your research project can be extrapolated (or transported) to a wider context. Such an extrapolation answers the 'So what?' question in relation to your research and completes the agenda that good research is for me, for us and for them, that is, it integrates first, second and third person research. This section is answering the question, 'Why should anyone who has not been involved directly in my research be interested in it?'

Action research projects are situation specific, and do not aim to create universal knowledge. At the same time, extrapolation from a local situation to more general situations is important. As an action researcher you are not claiming that every organization will behave as the one you have studied. But you can focus on some significant factors, consideration of which is useful for other organizations, perhaps those undergoing similar types of change processes.

For readers whose action research is directed at both a practical organizational outcome and an academic assessment, it may be useful for you to produce two documents. Organizational readers tend to be interested primarily in the story and its analysis, and be less interested in academic citations, critiques of methodology, literature reviews and discussions of theoretical differences between schools of thought, which are central to an academic dissertation. For the organizational readers, the researcher may produce a report which contains the core story and its analysis, omitting the academic requirements.

Dissemination

Dissemination of action research occurs in ways similar to all forms of research dissemination. A dissertation is held in a library, with its abstract circulated on abstract indexes. Articles and papers may be submitted to journals and books. Throughout this book, we have referenced many such published accounts.

Political sensitivities are typically critical issues in the writing of the research report and its dissemination. The content of the report may contain classified material or data of interest to competitors. Individual actors may be identifiable and their reported role in the events of the story may not be complimentary. Conventions relating to disguising the identity of the case and the actors may be applied. Krim (1988) and Moore (2007) do not identify the organizations in which their research took place, and probably, for most readers, their identity is irrelevant and subordinate to the theme and methodology of the research. Williander and Styhre (2006) and Roth et al. (2007) identify their organizations as Volvo and AstraZeneca respectively.

A particular application of action research dissemination involves those who have participated as co-researchers in the project. You may have a moral obligation to involve them in your formulation of conclusions. Meehan, when he had drafted his thesis, circulated a copy of his story and reflection chapters to the group and incorporated their comments. Subsequently, when he was writing his article with Coghlan, he circulated a draft to the group (Meehan and Coghlan, 2004).

Nielsen and Repstad (1993) advise:

- do not promise greater anonymity than you can keep
- take actors' analyses seriously
- point to potential solutions
- take the opportunity to discuss with as many as possible.

Publishing

From the field of action learning, Mann and Clarke (2007) have articulated a provocative slogan for practitioners who wish to publish accounts of their practice. 'Writing it down – writing it out – writing it up' captures an insightful approach which is a useful construct for reflecting on what you might want to say in an article out of your action research work and how you might say it. While the article uses action learning terminology, its core message of working through what your work is about, what it is really about and what it is really, really about is stimulating and enabling.

One particular dilemma that action researchers face in publishing their work is whether to use the personal first person or the impersonal third person narrative style when referring to themselves as the researcher or author. There is no consensus. Krim (1988), Holian (1999), Humphrey (2007) and Moore (2007), for example, use the personal 'I' throughout. Williander and Roth, as they are writing in collaboration, use the impersonal third person form (Williander and Styhre, 2006; Roth et al., 2007). A useful guideline in our experience is that if the report contains extensive reflection on the personal learning of the author-researcher as agent of the action in the story (as instanced by Krim, Humphrey and Moore), then the first person narrative adds considerable strength to the published report. Third person narrative gives a sense of objectivity and remove, while using the first person demands that the distinction between the story's narrative and the researcher's interpretation and sense making be very clearly distinct.

Marshall (2008) reflects on the role of form in her action research writing and suggests the following practices:

- accept and seek to express what is rather than what should be
- employ disciplines and respect emergence
- invoke the writer in your and our own direct voice whatever shape it takes
- create resonant spaces and conditions for writing
- defend emergent form and claim authority
- value the imaginal and metaphorical as guides to form.

Conclusions

Writing the action research dissertation is an act of learning and is itself an action research project. During the writing of the dissertation you draw together the complexities of all your experiences, your insights and your judgements – data which integrate your own personal learning as well as what took place in the system in which you work. Telling the story, making sense of it and applying a rigorous methodology to that sense making are directed toward the generation of useful knowledge which must produce outcomes which are of value to others.

TWELVE

In Conclusion

In this book we have explored a subject which continues to represent very common practice in post-experience master's and doctoral programmes within business education, healthcare, social work and third sector organizations.

Action research is about undertaking action and studying that action as it takes place. It is about improving practice through intervention, and demands rigorous preparation, planning, action, attention to process, reflection, replanning and validating claims to learning and theory generation. It is collaborative, involving interacting with others. We have focused on the dynamics of how this might be done when the action research is being undertaken in the organization of the action researcher, who is both aiming at achieving personal goals from the project and contributing to the organization. We drew on the image, adapted from Weisbord, of the actor-director in the act of making a film. Film making involves creating a script whereby the story of the movie is enacted by people in interaction with one another over time. The actor-director engages in both acting in the film and standing back to study how the shots are being taken, how the actors are performing and how subsequent shots need to be set up.

Action research operates within the realm of practical knowing, where the concern is practical outcomes and where what is familiar in one setting may be unfamiliar in another. Accordingly, our knowing in everyday situations is always incomplete and can only be completed by attending to the particular tasks and situations in which we are at a given time. A remembered set of insights is only approximately appropriate to the new situation. They are insights into situations which are similar but not identical. No two situations are identical, as time has passed, other events have taken place and we remember differently. This is why we reason, reflect and judge in a practical pattern of knowing in order to move from one setting to another, grasping what modifications are needed.

We began our exploration by locating action in the three core foundational elements: (1) how action research works through the spirals of cycles of constructing, planning, taking and evaluating action, and the meta cycle of content, process and premise reflection, (2) how the individual researcher learns in action and (3) how the whole process is genuinely collaborative. In all this the critical skills are introspective reflection and, when engaging with others, combining advocacy with inquiry. While many of the issues of any research project pertain to researching your own organization, the particular challenges posed by researching your own organization are around managing the closeness–distance tension in preunderstanding and in managing role duality, and around managing the politics in order that you have a future in the organization when the research is completed.

Engaging in insider action research poses serious challenges to first person inquiry for you as an insider action researcher. You are confronted primarily with the familiar – the people and organizational rituals that you encounter on a daily basis. As familiarity inhibits inquiry, the challenge is to develop ways of knowing what is familiar and to develop practical knowledge that enables you to understand what is taking place and to act in the shifting situations of organizational life. Inquiring and acting within the familiar evokes issues of role – the existing organizational roles and the adopted researcher role – and the accompanying challenges to hold both sets of roles concurrently and manage the tension between them. All such activity is political, so managing the political dynamics which pertain to being a change agent in an organization is critical for survival and success.

We have invited you to bring the general empirical method to your work. In essence, this method shows us that there are four basic, unfolding, and interrelated operations that fit into a pattern that characterizes the human, namely: experience of data, the effort to understand them, the judgement after relating the insight back to the evidence of data, and the decision about what to do about it. In first person terms, you are starting with yourself and the dynamic structure of knowing that is integral to being human. Once you understand how you know, you can transpose that dynamic operational structure into the basic, normative patterns of insider action research. Attention to experience, understanding and judgement, which leads to action, provides a methodology through which you can affirm what and how you know. Quality and rigour can be formulated by the notion of authenticity which is characterized by four process imperatives: be attentive (to the data); be intelligent (in inquiry); be reasonable (in making judgements); be responsible (in making decisions and taking action). The quest for authenticity for insider action researchers, then, is that struggle to engage in being attentive, intelligent, reasonable and responsible in confronting the challenges of preunderstanding, role duality and organizational politics (Table 12.1).

What you can do in an action research project in your own organization depends on the formal and psychological contract you have with the system on its commitment to learning in action juxtaposed with yours. This is critical to the nature of the project you can undertake. You draw on your preunderstanding of

Table 12.1 Authenticity

Process imperatives	Preunderstanding	Role duality	Organizational politics
Be attentive	What is familiar, how is it different when the ritual is the same	Role expectations and ambiguity/conflict	Own questions and dilemmas in acting politically Political behaviour of others
Be intelligent	Evidence you are being presented with as you work with the familiar	Understand the structural nature of roles	Understand the structural pattern of political action
Be reasonable	Weigh the evidence, distinguish inferences	Weigh evidence and judge conflicting role and demands	Weigh evidence and judge demands of political action
Be responsible	Take responsibility for what you do	Take responsibility for what you do	Take responsibility for what you do

the organization and how you can manage the twin roles of your regular organizational role and the researcher role. With that foundation we explored some important and useful issues: managing organizational politics, framing and selecting a project, implementing it and, finally, writing an action research dissertation.

In the past we have been asked questions such as: 'Given that doing insider research seems to be so complex and can be fraught with danger, wouldn't it be better to advise people not to select this approach as a means to do any academic research, particularly a major thesis for a postgraduate qualification?' In our view, while it can be difficult at times, the challenges that have to be overcome are usually well worth the effort. Intending insider researchers do need to be aware of the issues they may face, both to help prepare them to address these when they arise, and to know that this is a 'normal' part of the process. Wirth reflects:

> This research has been the most exciting and challenging learning process I have experienced professionally. It changed the way I view teaching and instruction, and it certainly changed my views on collaboration. The research has made me realize how much more we need to foster collaboration within the school system. (Ravitch and Wirth, 2007: 81)

Nielsen and Repstad (1993) describe the notion of researching your own organization in terms of a journey from nearness to distance and back and provide some practical advice. In terms of maintaining distance, they advise that you be aware of your preconceived ideas and prejudices about the organization and find rational theories to explain the organization as a way of distancing yourself. They suggest that you perform the role of devil's advocate by finding alternative theories which contradict the rational theories you have selected to explain the organization by engaging in dialogue with other members of the organization. In their view, you need to consider seriously the prevailing explanation of organizational problems, which might reflect analyses of symptoms. These would typically be: (1) the scapegoat syndrome, 'who is to blame'; (2) too little information, 'if only we had more'; and (3) too few resources, 'if only we had more'.

For some readers, doing action research in their own organizations is an exciting, demanding and invigorating prospect which will contribute considerably to their own learning and help their career development. For others it may seem daunting, with a high potential for self-destruction. In the words of Brookfield (1994), it smacks of impostorship, cultural suicide, lost innocence, road running and community. Can you survive doing action research in your own organization? Shepard (1997) provides a few rules of thumb for change agents which are also useful for insider action researchers:

- *Stay alive.* Care for yourself. Keep a life outside the project so as to be able to turn yourself on and off. Stay in touch with the purpose of the project and go with the flow.
- *Start where the system is.* Have empathy with the system and the people in it, particularly as it won't like being 'diagnosed'.
- *Never work uphill.* Keep working at collaboration and work in the most promising arena.
- *Innovation requires a good idea, initiative and a few friends.* Find the people who are ready to work on the project and get them working together.
- *Load experiments with success.* Work at building success steps along the way.
- *Light many fires.* Remember the notion of systems. Any part of a system is the way it is because of how the rest of the system is. As you work towards change in one part, other parts will push the system back to the way it was. Understand the interdependencies among subsystems and keep movement going in as many of them as you can.
- *Keep an optimistic bias.* Stay focused on vision and desired outcomes.
- *Capture the moment.* Stay in tune with yourself and the situation.

Friedman (2001) suggests four attributes: be proactive and reflective, be critical and committed, be independent and work well with others, and have aspirations and be realistic about limits. With these in mind we are confident that not only can you survive doing action research in your own organization but you can flourish and be successful.

We close with David's words:

I understand the subversive nature of action research and I like it. I like the counter-culturalness to it all. It seems all at once to be natural and unnatural, something so right and primal, so comforting and familiar, yet something that seems to have been taken away from us in some way. We have learned to be inactive, even in the face of injustice, and we are aware of it. We know it, yet we continue to be inactive. Action research, even in my little context in my little school in the suburbs, was in some way revolutionary, for it allowed people to feel more in control of their work lives, to be more in control of their day-to-day being, to be more connected with their reality that they experience every day. Something else I like about action research. I did not have the assistants journal, take notes, read books or participate in any activity other than those that helped them become better at what they do. We all knew in our hearts and souls that we were making our present situations better by working together and that is one very good reason to use action research.

References

Adler, N., Shani, A.B. (Rami) and Styhre, A. (2004) *Collaborative Research in Organizations*. Thousand Oaks, CA: Sage.

Adler, P.A. and Adler, P. (1987) *Membership Roles in Field Research*. Thousand Oaks, CA: Sage.

Alvesson, M. (2003) 'Methodology for close up studies: struggling with closeness and closure', *Higher Education*, 46: 167–93.

Alvesson, M. and Skoldberg, K. (2000) *Reflexive Methodology: New Vistas for Qualitative Research*. Thousand Oaks, CA: Sage.

Anderson, G., Herr, K. and Nihlen, A. (1994) *Studying Your Own School*. Thousand Oaks, CA: Corwin.

Anderson, V. and Johnson, L. (1997) *Systems Thinking Basics: From Concepts to Causal Loops*. Cambridge, MA: Pegasus.

Argyris, C. (1993) *Knowledge for Action*. San Francisco: Jossey-Bass.

Argyris, C. (2003) 'Actionable knowledge', in T. Tsoukas and C. Knudsen (eds), *The Oxford Handbook of Organization Theory*. Oxford: Oxford University Press. pp. 423–52.

Argyris, C. (2004) *Reasons and Rationalizations: The Limits to Organizational Knowledge*. New York: Oxford University Press.

Argyris, C. and Schon, D. (1996) *Organizational Learning II*. Reading, MA: Addison-Wesley.

Argyris, C., Putnam, R. and Smith, D. (1985) *Action Science*. San Francisco: Jossey-Bass.

Ashforth, B., Kreiner, G. and Fugate, M. (2000) 'All in a day's work: boundaries and micro role transitions', *Academy of Management Review*, 25 (3): 472–91.

Bargal, D. (2006) 'Personal and intellectual influences leading to Lewin's paradigm of action research', *Action Research*, 4 (4): 367–88.

Bartunek, J.M. (2003) *Organizational and Educational Change: The Life and Role of a Change Agent Group*. Mahwah, NJ: Erlbaum.

Bartunek, J.M. (2008) 'Insider/outsider team research: the development of the approach and its meanings', in A.B. (Rami) Shani, S.A. Mohrman, W. Pasmore, B. Stymne and N. Adler (eds), *Handbook of Collaborative Management Research*. Thousand Oaks, CA: Sage. pp. 73–92.

Bartunek, J.M. and Louis, M.R. (1996) *Insider/Outsider Team Research*. Thousand Oaks, CA: Sage.

Bartunek, J.M., Crosta, T.E., Dame, R.F. and LeLacheur, D.F. (2000) 'Managers and project leaders conducting their own action research interventions', in R.T. Golembiewski (ed.), *Handbook of Organizational Consultation*, 2nd edn. New York: Dekker. pp. 59–70.

Baskerville, R. and Pries-Heje, J. (1999) 'Grounded action research: a method for understanding IT in practice', *Accounting Management and Information Technology*, 9: 1–23.

Bateson, G. (1972) *Steps to an Ecology of Mind*. San Francisco: Ballantine.

Beckhard, R. (1997) *Agent of Change, My Life, My Work*. San Francisco: Jossey-Bass.

Beckhard, R. and Harris, R. (1987) *Organizational Transitions: Managing Complex Change*, 2nd edn. Reading, MA: Addison-Wesley.

Beckhard, R. and Pritchard, W. (1992) *Changing the Essence: The Art of Creating and Leading Fundamental Change in Organizations*. San Francisco: Jossey-Bass.

Bell, S. (1998) 'Self-reflection and vulnerability in action research: bringing forth new worlds in our learning', *Systemic Practice and Action Research*, 11 (2): 179–91.

Benne, K.D. (1959) 'Some ethical problems in group and organizational consultation', *Journal of Social Issues*, 15 (2): 60–7.

Bentz, V.M. and Shapiro, J.J. (1998) *Mindful Inquiry in Social Research*. Thousand Oaks, CA: Sage.

Bjorkman, H. and Sundgren, M. (2005) 'Political entrepreneurship in action research: learning from two cases', *Journal of Organizational Change Management*, 18 (5): 399–415.

Bolman, D. and Deal, T. (2008) *Reframing Organizations*, 3rd edn. San Francisco: Jossey-Bass.

Boser, S. (2006) 'Ethics and power in community–campus partnerships for research', *Action Research*, 4 (1): 9–22.

Boud, D. (2001) 'Using journal writing to enhance reflective practice', in L.M. English and M.A. Gillen (eds), *Promoting Journal Writing in Adult Education*. San Francisco: Jossey-Bass. pp. 9–18.

Boud, D., Keogh, R. and Walker, D. (1985) *Reflection: Turning Experience into Learning*. Abingdon: Routledge.

Bourdieu, P. (1990) *The Logic of Practice*. Cambridge: Polity.

Bourner, T. (2003) 'Assessing reflective learning', *Education and Training*, 45 (5): 267–72.

Bradbury, H., Mirvis, P., Neilsen, E. and Pasmore, W. (2008) 'Action research at work: creating the future following the path from Lewin', in P. Reason and H. Bradbury (eds), *Handbook of Action Research*, 2nd edn. London: Sage. pp. 77–92.

Brannick, T. and Coghlan, D. (2007) 'In defense of being "native": the case for insider academic research', *Organizational Research Methods*, 10 (1): 59–74.

Brookfield, S. (1994) 'Tales from the dark side: a phenomenography of adult critical reflection', *International Journal of Lifelong Education*, 13 (3): 203–16.

Brydon-Miller, M. (2008) 'Ethics and action research: deepening our commitment to social justice and redefining systems of democratic practice', in P. Reason and H. Bradbury (eds), *Handbook of Action Research*, 2nd edn. London: Sage. pp. 199–210.

Brydon-Miller, M. and Greenwood, D. (2006) 'A re-examination of the relationship between action research and human subjects review process', *Action Research*, 4 (1): 117–28.

Brydon-Miller, M., Greenwood, D. and Eikeland, O. (guest eds) (2006) Special Issue: Ethics and Action Research. *Action Research*, 4 (1).

Brydon-Miller, M., Greenwood, D. and Maguire, P. (2003) 'Why action research?', *Action Research*, 1 (1): 9–28.

Buchanan, D. and Badham, R. (2008) *Power, Politics and Organizational Change: Winning the Turf Game*, 2nd edn. London: Sage.

Buchanan, D. and Boddy, D. (1992) *The Expertise of the Change Agent*. London: Prentice-Hall.

Bunker, B. and Alban, B. (2006) *The Handbook of Large Group Methods*. San Francisco: Jossey-Bass.

Buono, A.F. and Kerber, K.W. (2008) 'The challenges of organizational change: enhancing organizational change capacity', *Revue Sciences de Gestion*, 65: 99–118.

Buono, A.F. and Savall, H. (2007) *Socio-economic Interventions in Organizations*. Charlotte, NC: Information Age.

Burke, W.W. (2008) *Organization Change: Theory and Practice*, 2nd edn. Thousand Oaks, CA: Sage.

Burnes, B. (2007) 'Kurt Lewin and the Harwood Studies: the foundations of OD', *Journal of Applied Behavioral Science*, 43 (2): 213–31.

Burns, D. (2007) *Systemic Action Research*. Bristol: Polity.

Bushe, G.R. and Marshak, R.J. (2008) 'The postmodern turn in OD', *OD Practitioner*, 40 (4): 9–11.

Campbell, D. (2000) *The Socially Constructed Organization*. London: Karnac.

Cassell, C. and Johnson, P. (2006) 'Action research: explaining the diversity', *Human Relations*, 59 (6): 783–814.

Chandler, D. and Torbert, W.R. (2003) 'Transforming inquiry in action: interweaving 27 flavors of action research', *Action Research*, 1 (2): 133–52.

Coch, L. and French, J.R.P. (1948) 'Overcoming resistance to change', *Human Relations*, (1): 512–32.

Coghlan, D. (1993) 'Learning from emotions through journaling', *Journal of Management Education*, 17 (1): 90–4.

Coghlan, D. (2002) 'Interlevel dynamics in systemic action research', *Systemic Practice and Action Research*, 15 (4): 273–83.

Coghlan, D. (2003) 'Practitioner research for organizational knowledge: mechanistic- and organistic-oriented approaches to insider action research', *Management Learning*, 34 (4): 451–63.

Coghlan, D. (2007) 'Insider action research doctorates: generating actionable knowledge', *Higher Education*, 54: 293–306.

Coghlan, D. (2008a) 'Exploring insight: the role of insight in a general empirical method in action research for organization change and development', *Revue Sciences de Gestion*, 65: 343–55.

Coghlan, D. (2008b) 'Authenticity as first person practice: an exploration based on Bernard Lonergan', *Action Research*, 6 (3): 339–43.

Coghlan, D. (2009) 'Toward a philosophy of clinical inquiry/research,' *Journal of Applied Behavioral Science*, 45 (1): 106–21.

Coghlan, D. and Casey, M. (2001) 'Action research from the inside: issues and challenges in doing action research in your own hospital', *Journal of Advanced Nursing*, 35: 674–82.

Coghlan, D. and Coughlan, P. (2005) 'Collaborative research across borders and boundaries: action research insights from the *CO-IMPROVE* project', in R. Woodman and W. Pasmore (eds), *Research in Organizational Change and Development*, Vol. 15. Greenwich, CT: JAI. pp. 275–95.

Coghlan, D. and Holian, R. (guest eds) (2007) Special Issue: Insider Action Research. *Action Research*, 5 (1).

Coghlan, D. and Jacobs, C. (2005) 'Kurt Lewin on reeducation: foundations for action research', *Journal of Applied Behavioral Science*, 41 (4): 444–57.

Coghlan, D. and Pedler, M. (2006) 'Action learning dissertations: structure, supervision and examination', *Action Learning: Research and Practice*, 3 (2): 127–40.

Coghlan, D. and Rashford, N.S. (1990) 'Uncovering and dealing with organizational distortions', *Journal of Managerial Psychology*, 5 (3): 17–21.

Coghlan, D. and Rashford, N.S. (2006) *Organization Change and Strategy: An Interlevels Dynamics Approach*. Abingdon: Routledge.

Coghlan, D. and Shani, A.B. (Rami) (2005) 'Roles, politics and ethics in action research design', *Systemic Practice and Action Research*, 18 (6): 533–46.

Coghlan, D. and Shani, A.B. (Rami) (2008) 'Collaborative management research through communities of inquiry', in A.B. (Rami) Shani, S.A. Mohrman, W. Pasmore, B. Stymne and N. Adler (eds), *Handbook of Collaborative Management Research*. Thousand Oaks, CA: Sage. pp. 601–14.

Coghlan, D., Dromgoole, T., Joynt, P. and Sorensen, P. (2004) *Managers Learning in Action: Research, Learning and Education*. London: Routledge.

Cooklin, A. (ed.) (1999) *Changing Organizations: Clinicians as Agents of Change*. London: Karnac.

Cooperrider, D.L. and Srivastva, S. (1987) 'Appreciative inquiry in organizational life', in R. Woodman and W. Pasmore (eds), *Research in Organizational Change and Development*, Vol. 1. Greenwich, CT: JAI. pp. 129–69.

Cooperrider, D.L. and Whitney, D. (2005) *Appreciative Inquiry: A Positive Revolution in Change*. San Francisco: Berrett-Koehler.

Cornell, A.W. (1996) *The Power of Focusing*. Oakland, CA: New Harbinger.

Coughlan, P. and Coghlan, D. (2009) 'Action research', in C. Karlsson (ed.), *Researching Operations Management*. New York: Routledge. pp. 236–64.

Coyer, F., Courtney, M. and O'Sullivan, J. (2007) 'Establishing an action research group to explore family-focused nursing in the intensive care unit', *International Journal of Nursing Practice*, 13: 14–23.

Darling, M. and Parry, C. (2000) *From Post-Mortem to Living Practice: An In-Depth Study of the Evolution of the After Action Review*. Boston: Signet.

Deane, C. (2004) 'Learning to change', in D. Coghlan, T. Dromgoole, P. Joynt and P. Sorensen (eds), *Managers Learning in Action: Research, Learning and Education*. London: Routledge. pp. 9–23.

De Guerre, D. (2002) 'Doing action research in one's own organization: an ongoing conversation over time', *Systemic Practice and Action Research*, 15 (4): 331–49.

DeTardo-Bora, K. (2004) 'Action research in a world of positivist-oriented review boards', *Action Research*, 2 (3): 237–53.

De Vos, H. (1987) 'Common sense and scientific thinking', in F. van Hoolthoon and D.R. Olsen (eds), *Common Sense: The Foundation for Social Science*. Lanham, MD: University Press of America. pp. 345–59.

Dewing, J. (2009) 'Prioritising mealtime care, patient choice, and nutritional assessment were important for older inpatients' mealtime experiences', *Evidence-Based Nursing*, 12: 30–40.

Dick, B. (1999) 'You want to do an action research thesis?' http://www.scu.edu.au/schools/sawd/arr/arth/arthesis.html.

Dickens, L. and Watkins, K. (1999) 'Action research: rethinking Lewin', *Management Learning*, 30 (2): 127–40.

Dilworth, L. and Willis, V. (2003) *Action Learning: Images and Pathways*. Malabar, FL: Krieger.

Docherty, P., Ljung, A. and Stjernberg, T. (2006) 'The changing practice of action research', in J. Lowstedt and T. Stjernberg (eds), *Producing Management Knowledge*. Abingdon: Routledge. pp. 221–36.

Docherty, P., Kira, M. and Shani, A.B. (Rami) (2008) *Sustainable Work Systems: Past, Present and Future of Social Sustainability*. Abingdon: Routledge.

Docherty, P., Kira, M. and Shani, A.B. (Rami) (2009) 'Organizational development for social sustainability in work systems', in W. Pasmore, R. Woodman and A.R. (Rami) Shani (eds), *Research in Organizational Change and Development*, Vol. 17. Bradford: Emerald. pp. 77–144.

Dutton, J.E. and Ottensmeyer, E. (1987) 'Strategic issues management systems: forms, functions and contexts', *Academy of Management Review*, 12 (2): 355–65.

Dutton, J.E., Fahey, L. and Narayanan, V.K. (1983) 'Toward understanding strategic issue diagnosis', *Strategic Management Journal*, 4: 307–23.

Eden, C. and Huxham, C. (1996) 'Action research for the study of organizations', in S.R. Clegg, C. Hardy and W.R. Nord (eds), *Handbook of Organization Studies*. Thousand Oaks, CA: Sage. pp. 526–42.

Eikeland, O. (2006a) 'Phronesis, Aristotle and action research', *International Journal of Action Research*, 2 (1): 5–53.

Eikeland, O. (2006b) 'Condescending ethics and action research', *Action Research*, 4 (1): 37–48.

Eikeland, O. (2008) *The Ways of Aristotle: Aristotelian Phronesis, Aristotelian Philosophy of Dialogue and Action Research*. Bern: Lang.

Elizur, Y. (1999) '"Inside" consultation through self-differentiation: stimulating organization development in the IDF's care of intractable, war-related, traumatic disorders', in A. Cooklin (ed.), *Changing Organizations: Clinicians as Agents of Change*. London: Karnac. pp. 141–68.

Evans, M. (1997) 'An action research enquiry into reflection in action as part of my role as a deputy headteacher'. PhD thesis, Kingston University, UK. http://www.bath.ac.uk/-edsajw/moyra.html.

Evered, R. M. and Louis, M.R. (1981) 'Alternatives perspectives in the organizational sciences: "inquiry from the inside" and "inquiry from the outside"', *Academy of Management Review*, 6: 385–95.

Ferguson, P. and Ferguson, B. (2001) 'Shooting oneself in the foot: an investigation of some issues in conducting insider research', paper presented at 24th HERDSA Conference, Newcastle, Australia.

Fisher, D., Rooke, D. and Torbert, B. (2000) *Personal and Organizational Transformations through Action Inquiry*. Boston: Edge\Work.

Fisher, R. and Sharp, A. (1998) *Getting It Done*. New York: Harper.

Fisher, R. and Ury, W. (1986) *Getting to Yes*. London: Business Books.

Flanagan, J. (1997) *Quest for Self-Knowledge: An Essay in Lonergan's Philosophy*. Toronto: University of Toronto Press.

Flyvbjerg, B. (2001) *Making Social Science Matter*. Cambridge: Cambridge University Press.

Foster, M. (1972) 'An introduction to the theory and practice of action research in work organizations', *Human Relations*, 25 (6): 529–56.

French, W. and Bell, C. (1999) *Organization Development: Behavioral Science Interventions for Organization Improvement*, 6th edn. Englewood Cliffs, NJ: Prentice-Hall.

Fricke, W. and Totterdill, R. (2004) *Action Research in Workplace Innovation and Regional Development*. Amsterdam: Benjamins.

Friedman, V. (2001) 'The individual as agent of organizational learning', in M. Dierkes, J. Child, I. Nonaka and A. Berthoin Antal (eds), *Handbook of Organizational Learning*. Oxford: Oxford University Press. pp. 398–414.

Friedman, V. and Rogers, T. (2008) 'Action science: linking causal action and meaning making in action research', in P. Reason and H. Bradbury (eds), *Handbook of Action Research*, 2nd edn. London: Sage. pp. 252–65.

Frost, P. and Robinson, S. (1999) 'The toxic handler: organizational hero – and casualty', *Harvard Business Review*, July–August: 97–106.

Gellerman, W., Frankel, M. and Ladenson, R. (1990) *Values and Ethics in Organization and Human System Development*. San Francisco: Jossey-Bass.

Gendlin, E. (1981) *Focusing*. New York: Bantam.

Gergen, K. and Gergen, M. (2008) 'Social construction and research on action', in P. Reason and H. Bradbury (eds), *Handbook of Action Research,* 2nd edn. London: Sage. pp. 159–71.

Gibbons, M., Limoges, C., Nowotny, H., Schwartzman, S., Scott, P. and Trow, M. (1994) *The New Production of Knowledge*. London: Sage.

Goffman, E. (1959) *The Presentation of Self in Everyday Life*. New York: Doubleday.

Goode, L.M. and Bartunek, J.M. (1990) 'Action research in an underbounded setting', *Consultation*, 9 (3): 209–28.

Gorinski, R. and Ferguson, P. (1997) '(Ex)changing experiences of insider research', paper presented at NZARE Conference, Auckland, NZ.

Greenwood, D. and Levin, M. (2007) *Introduction to Action Research*, 2nd edn. Thousand Oaks, CA: Sage.

Guba, E. and Lincoln, Y. (1994) 'Competing paradigms in qualitative research', in N.K. Denzin and Y. Lincoln (eds), *Handbook of Qualitative Research*. Thousand Oaks, CA: Sage. pp. 93–9.

Gummesson, E. (2000) *Qualitative Methods in Management Research*, 2nd edn. Thousand Oaks, CA: Sage.

Gustavsen, B. (2003) 'New forms of knowledge production and the role of action research', *Action Research*, 1: 153–64.

Gustavsen, B. and Englestad, P. (1986) 'The design of conferences and the evolving role of democratic dialogue in changing working life', *Human Relations*, 39: 101–16.

Harrison, M. (2005) *Diagnosing Organizations: Methods, Models and Processes*, 3rd edn. Thousand Oaks, CA: Sage.

Harrison, M. and Shirom, A. (1999) *Organizational Diagnosis and Assessment: Bridging Theory and Practice*. Thousand Oaks, CA: Sage.

Haslebo, G. and Nielsen, K.S. (2000) *Systems and Meaning: Consulting in Organizations*. London: Karnac.

Hatchuel, A. and David, A. (2008) 'Collaborating for management research: from action research to intervention research in management', in A.B. (Rami) Shani, S.A. Mohrman, W. Pasmore,

B. Stymne and N. Adler (eds), *Handbook of Collaborative Management Research.* Thousand Oaks, CA: Sage. pp. 143–61.

Heron, J. (1996) *Cooperative Inquiry.* London: Sage.

Heron, J. and Reason, P. (2008) 'Extending epistemology with a co-operative inquiry', in P. Reason and H. Bradbury (eds), *Handbook of Action Research,* 2nd edn. London: Sage. pp. 367–80.

Herr, K. and Anderson, G. (2005) *The Action Research Dissertation.* Thousand Oaks, CA: Sage.

Hill, M.R. (1993) *Archival Strategies and Techniques.* Thousand Oaks, CA: Sage.

Hilsen, A.I. (2006) 'And they shall be known by their deeds: ethics and politics in action research', *Action Research,* 4 (1): 23–36.

Hockley, J. and Froggatt, K. (2006) 'The development of palliative care knowledge in care homes for older people: the place of action research', *Palliative Medicine,* 20: 835–43.

Holian, R. (1999) 'Doing action research in my own organization: ethical dilemmas, hopes and triumphs', *Action Research International,* paper 3. http://www.scu.edu.au/schools/sawd/ari/ari/holian.html.

Holian, R. and Brooks, R. (2004) 'The Australian National Statement on Ethical Conduct in Research: application and implementation for "insider" applied research in business', *Action Research International,* Paper 7. http://scu/au/schools/gcm/ar/ari/p.rholian04.html.

Holman, P., Devane, T. and Cady, S. (2007) *The Change Handbook: Group Methods for Shaping the Future,* 2nd edn. San Francisco: Berrett-Koehler.

Homa, P. (1998) 'Re-engineering the Leicester Royal Infirmary healthcare process'. Unpublished PhD thesis, Henley Management College and Brunel University, UK. http://www.bath.ac.uk/-edsajw/erica.html.

Hughes, I. (2008) 'Action research in healthcare', in P. Reason and H. Bradbury (eds), *Handbook of Action Research,* 2nd edn. London: Sage. pp. 381–93.

Humphrey, C. (2007) 'Insider–outsider: activating the hyphen', *Action Research,* 5 (1): 11–26.

International Journal of Action Research, Special Issue, Vol. 3, Issues 1 and 2, 2007.

Johansson, A. and Lindult, E. (2008) 'Emancipation or workability? Critical versus pragmatic scientific orientation in action research', *Action Research,* 6 (1): 95–115.

Johnson, P. and Duberley, J. (2000) *Understanding Management Research.* London: Sage.

Jones, S.P., Auton, M.F., Burton, C.R and Watkins, C.L. (2008) 'Engaging service users in the development of stroke services: an action research study', *Journal of Clinical Nursing,* 17: 1270–9.

Kakabadse, A. (1991) 'Politics and ethics in action research', in N. Craig Smith and P. Dainty (eds), *The Management Research Handbook.* London: Routledge. pp. 289–99.

Kaplan, R.S. (1998) 'Innovation action research: creating new management theory and practice', *Journal of Management Accounting Research,* 10: 89–118.

Katz, D. and Kahn, R.L. (1978) *The Social Psychology of Organizations,* 2nd edn. New York: McGraw-Hill.

Kelman, H.C. (1965) 'Manipulation of human behavior: an ethical dilemma for the social scientist', *Journal of Social Issues,* 21 (2): 31–46.

Kleiner, A. and Roth, G. (1997) 'How to make experience your company's best teacher', *Harvard Business Review,* September–October: 172–77.

Kleiner, A. and Roth, G. (2000) *Oil Change.* New York: Oxford University Press.

Koch, N. (2007) *Information Systems and Action Research.* New York: Springer.

Koch, T. and Kralik, D. (2006) *Participatory Action Research in Healthcare.* Oxford: Blackwell.

Kolb, D. (1984) *Experiential Learning.* Englewood Cliffs, NJ: Prentice-Hall.

Krim, R. (1988) 'Managing to learn: action inquiry in city hall', in P. Reason (ed.), *Human Inquiry in Action.* London: Sage. pp. 144–62.

Levin, M. (2003) 'Action research and the research community', *Concepts and Transformation,* 8 (3): 275–80.

Lewin, K. (1946/1997) 'Action research and minority problems', in K. Lewin, *Resolving Social Conflicts: Selected Papers on Group Dynamics*. Ed. G. Lewin. Reprinted 1997, Washington, DC: American Psychological Association. pp. 144–54.

Lewin, K. (1948/1999) 'Group decision and social change'. Reprinted 1999 in M. Gold (ed.), *The Complete Social Scientist: A Kurt Lewin Reader*. Washington, DC: American Psychological Association. pp. 265–84.

Lincoln, Y. (2001) 'Engaging sympathies: relationships between action research and social constructivism', in P. Reason and H. Bradbury (eds), *Handbook of Action Research*. London: Sage. pp. 124–32.

Lippitt, R. (1961) 'Value-judgment problems of the social scientist in action research', in W. Bennis, K. Benne and R. Chin (eds), *The Planning of Change*. New York: Holt, Rinehart and Winston. pp. 689–94.

Lippitt, R. (1979) 'Kurt Lewin, action research and planned change', paper provided by the author.

Lonergan, B.J. (1992) *Insight: An Essay in Human Understanding. The Collected Works of Bernard Lonergan*, Vol. 3. Eds F. Crowe and R. Doran. Toronto: University of Toronto Press (original publication London: Longmans, 1957).

Ludema, J. and Fry, R. (2008) 'The practice of appreciative inquiry', in P. Reason and H. Bradbury (eds), *Handbook of Action Research*, 2nd edn. London: Sage. pp. 280–96.

Lykes, M.B. and Mallona, A. (2008) 'Towards transformational liberation: participatory and action research and praxis', in P. Reason and H. Bradbury (eds), *Handbook of Action Research,* 2nd edn. London: Sage. pp. 106–20.

Lynch, K. (1999) 'Equality studies, the academy and the role of research in emancipatory social change', *Economic and Social Review*, 30 (1): 41–69.

McArdle, K. and Reason, P. (2008) 'Organization development and action research', in T. Cummings (ed.), *Handbook of Organization Development*. Thousand Oaks, CA: Sage. pp. 123–36.

McCaughan, N. and Palmer, B. (1994) *Systems Thinking for Harassed Managers*. London: Karnac.

McGill, I. and Brockbank, A. (2004) *The Action Learning Handbook*. London: RoutledgeFalmer.

MacLean, D., MacIntosh, R. and Grant, S. (2002) 'Model 2 management research', *British Journal of Management*, 13 (2): 189–207.

McMullan, W. and Cahoon, A. (1979) 'Integrating abstract conceptualizing with experiential learning', *Academy of Management Review*, 4 (3): 453–8.

McNiff, J., Lomax, P. and Whitehead, J. (2003) *You and Your Action Research Project*, 2nd edn. London: Routledge.

McTaggart, R. (1998) 'Is validity really an issue in PAR?', *Studies in Culture, Organization and Societies*, 4 (2): 211–37.

Mann, P. and Clarke, D. (2007) 'Writing it down–writing it out–writing it up: researching our practice through action learning', *Action Learning: Research and Practice*, 4 (2): 153–71.

Marrow, A.J. (1969) *The Practical Theorist*. New York: Basic.

Marshall, J. (1995) *Women Managers Moving On*. London: Routledge.

Marshall, J. (1999) 'Living life as inquiry', *Systemic Practice and Action Research*, 12 (2): 155–71.

Marshall, J. (2001) 'Self-reflective inquiry practices', in P. Reason and H. Bradbury (eds), *Handbook of Action Research*. London: Sage. pp. 433–9.

Marshall, J. (2008) 'Finding form in writing for action research', in P. Reason and H. Bradbury (eds), *Handbook of Action Research*, 2nd edn. London: Sage. pp. 682–95.

Marshall, J. and Reason, P. (2007) 'Quality in research as taking an "attitude of inquiry"', *Management Research News*, 30 (5): 368–80.

Martin, A. (2008) 'Action research on a large scale: issues and practices', in P. Reason and H. Bradbury (eds), *Handbook of Action Research*, 2nd edn. London: Sage. pp. 394–406.

Meehan, C. and Coghlan, D. (2004) 'Managers as healing agents: a cooperative inquiry approach', *Systemic Practice and Action Research*, 17 (2): 407–23.

Melchin, K.R. and Picard, C.A. (2008) *Transforming Conflict through Insight*. Toronto: Toronto University Press.

Meyerson, D. (2001) *Tempered Radicals: How People Use Difference to Inspire Change at Work*. Boston: Harvard Business School Press.

Mezirow, J. (1991) *Transformative Dimensions of Adult Learning*, San Francisco: Jossey-Bass.

Mitki, Y., Shani, A.B. (Rami) and Stjernberg, T. (2000) 'A typology of change programs and their differences from a solid perspective', in R.T. Golembiewski (ed.), *Handbook of Organizational Consultation*, 2nd edn. New York: Dekker. pp. 777–85.

Moon, J. (1999) *Learning Journals: A Handbook for Academics, Students and Professional Development*: London: Kogan Page.

Moore, B. (2007) 'Original sin and action research', *Action Research*, 5 (1): 27–39.

Morton, A. (1999) 'Ethics in action research', *Systemic Practice and Action Research*, 12 (2): 219–22.

Nadler, D.A. (1977) *Feedback and Organization Development: Using Data-Based Methods*. Reading, MA: Addison-Wesley.

Nadler, D.A. (1998) *Champions of Change*. San Francisco: Jossey-Bass.

Neilsen, E. (2006) 'But let us not forget John Collier: commentary on David Bargal's personal and intellectual influences leading to Lewin's paradigm of action research', *Action Research*, 4 (4): 389–99.

Nielsen, J.C.R. and Repstad, P. (1993) 'From nearness to distance – and back: analyzing your own organization', *Copenhagen Business School, Institute of Organizational and Industrial Sociology*, papers in organizations 14.

Nielsen, K.A. and Svensson, L. (2006) *Action Research and Interactive Research: Beyond Theory and Practice*. Maastricht: Shaker.

Pedler, M. (2008) *Action Learning for Managers*, 2nd edn. London: Gower.

Pedler, M. and Burgoyne, J. (2008) 'Action learning', in P. Reason and H. Bradbury (eds), *Handbook of Action Research*. London: Sage. pp. 319–32.

Peter, J.P. and Olsen, J.C. (1983) 'Is marketing science?', *Journal of Marketing*, 47: 111–25.

Pettigrew, P. (2003) 'Power, conflicts and resolutions: a change agent's perspective on conducting action research within a multiorganizational partnership', *Systemic Practice and Action Research*, 16 (6): 375–91.

Phillips, J.L., Davidson, P.M., Jackson, D. and Kristjanson, L.J. (2008) 'Multi-faceted palliative care intervention: aged care nurses' and care assistants' perceptions and experiences', *Journal of Advanced Nursing*, 62: 216–27.

Pine, G.J. (2008) *Teacher Action Research*. Thousand Oaks, CA: Sage.

Portillo, M.C. (2009) 'Understanding the practical and theoretical development of social rehabilitation through action research', *Journal of Clinical Nursing*, 18: 234–45.

Preskill, H. and Torres, R.T. (1999) *Evaluative Inquiry for Learning in Organizations*. Thousand Oaks, CA: Sage.

Punch, M. (1994) 'Politics and ethics in qualitative research', in N.K. Denzin and Y.S. Lincoln (eds), *Handbook of Qualitative Research*. Thousand Oaks, CA: Sage. pp. 83–97.

Putnam, R. (1991) 'Recipes and reflective learning', in D. Schon (ed.), *The Reflective Turn: Case Studies in and on Educational Practice*. New York: Teachers' College of Columbia Press. pp. 145–63.

Quick, J., Paulus, P., Whittington, J., Larey, T. and Nelson, D. (1996) *Management Development, Well-Being and Health*. Chichester: Wiley.

Raelin, J.A. (1999) 'Preface', *Management Learning*, 30 (2): 115–25.

Raelin, J.A. (2008) *Work-Based Learning: Bridging Knowledge and Action in the Workplace*, rev. edn. San Francisco: Jossey-Bass.

Rahman, M.A. (2008) 'Some trends in the praxis of participatory action research', in P. Reason and H. Bradbury (eds), *Handbook of Action Research*. London: Sage. pp. 49–62.

Ramirez, I. and Bartunek, J.M. (1989) 'The multiple realities and experience of internal organization development consultation in health care', *Journal of Organizational Change Management*, 2 (1): 40–56.

Ravitch, S.M. and Wirth, S. (2007) 'Developing a pedagogy of opportunity for students and their teachers', *Action Research*, 5 (1): 75–91.

Reason, P. (1988) *Human Inquiry in Action*. London: Sage.

Reason, P. (1999) 'Integrating action and reflection through cooperative inquiry', *Management Learning*, 30 (2): 207–26.

Reason, P. (2006) 'Choice and quality in action research practice', *Journal of Management Inquiry*, 15 (2): 187–203.

Reason, P. and Bradbury, H. (2008) *Handbook of Action Research*, 2nd edn. London: Sage.

Reason, P. and Marshall, J. (1987) 'Research as personal process', in D. Boud and V. Griffin (eds), *Appreciating Adult Learning*. London: Kogan Page. pp. 112–26.

Reason, P. and Torbert, W.R. (2001) 'The action turn: toward a transformational social science', *Concepts and Transformation*, 6 (1): 1–37.

Reed, J. (2007) *Appreciative Inquiry: Research for Change*. Thousand Oaks, CA: Sage.

Revans, R. (1982) *The Origins and Growth of Action Learning*. Bromley: Chartwell-Bratt.

Revans, R. (1998) *ABC of Action Learning*. London: Lemos and Crane.

Riemer, J. (1977) 'Varieties of opportunistic research', *Urban Life*, 5 (4): 467–77.

Riordan, P. (1995) 'The philosophy of action science', *Journal of Managerial Psychology*, 10 (6): 6–13.

Rogers, C.R. (1958) 'The characteristics of a helping relationship', *Personnel and Guidance Journal*, 37: 6–16.

Ross, R. (1994) 'The ladder of inference', in P. Senge, C. Roberts, R. Ross, B. Smith and A. Kleiner (eds), *The Fifth Discipline Fieldbook*. London: Brealey. pp. 242–6.

Ross, R. and Roberts, C. (1994) 'Balancing inquiry and advocacy', in P. Senge, C. Roberts, R. Ross, B. Smith and A. Kleiner (eds), *The Fifth Discipline Fieldbook*. London: Brealey. pp. 253–9.

Roth, G. and Bradbury, H. (2008) 'Learning history: an action research in support of actionable learning', in P. Reason and H. Bradbury (eds), *Handbook of Action Research*, 2nd edn. London: Sage. pp. 350–65.

Roth, G. and Kleiner, A. (2000) *Car Launch*. New York: Oxford University Press.

Roth, J., Sandberg, R. and Svensson, C. (2004) 'The dual role of the insider action researcher', in N. Adler, A.B. (Rami) Shani and A. Styhre (eds), *Collaborative Research in Organizations*. Thousand Oaks, CA: Sage. pp. 117–34.

Roth, J., Shani, A.B. (Rami) and Leary, M. (2007) 'Facing the challenges of new capability development within a biopharma company', *Action Research*, 5 (1): 41–60.

Rousseau, D. (1985) 'Issues in organizational research: multi-level and cross-level perspectives', in L.L. Cummings and B.M. Staw (eds), *Research in Organizational Behavior*, Vol. 7. Greenwich, CT: JAI. pp. 1–37.

Rowan, J. (2000) 'Research ethics', *International Journal of Psychotherapy*, 5 (2): 103–10.

Rudolph, J., Taylor, S. and Foldy, E. (2001) 'Collaborative off-line reflection: a way to develop skill in action science and action inquiry', in P. Reason and H. Bradbury (eds), *Handbook of Action Research*, 2nd edn. London: Sage. pp. 405–12.

Sagor, R. (2004) *The Action Research Guidebook*. Thousand Oaks, CA: Corwin.

Schein, E.H. (1987) *The Clinical Perspective in Fieldwork*. Newbury Park, CA: Sage.

Schein, E.H. (1995) 'Process consultation, action research and clinical inquiry: are they the same?', *Journal of Managerial Psychology*, 10 (6): 14–19.

Schein, E.H. (1996) 'Kurt Lewin's change theory in the field and in the classroom: notes toward a model of managed learning', *Systems Practice*, 9 (1): 27–48.

Schein, E.H. (1997) 'Organizational learning: what is new?', in M.A. Rahim, R.T. Golembiewski and L.E. Pate (eds), *Current Topics in Management*, Vol. 2. Greenwich, CT: JAI. pp. 11–25.

Schein, E.H. (1999) *Process Consultation Revisited: Building the Helping Relationship.* Reading, MA: Addison-Wesley.

Schein, E.H. (2003) *DEC is Dead, Long Live DEC.* San Francisco: Berrett-Koehler.

Schein, E.H. (2004) *Organizational Culture and Leadership*, 3rd edn. San Francisco: Jossey-Bass.

Schein, E.H. (2008) 'Clinical inquiry/research', in P. Reason and H. Bradbury (eds), *Handbook of Action Research*, 2nd edn. London: Sage. pp. 267–79.

Schein, E.H. (2009) *Helping.* San Francisco: Berrett-Koehler.

Schon, D. (1983) *The Reflective Practitioner.* New York: Basic.

Schon, D. (1987) *Educating the Reflective Practitioner.* San Francisco: Jossey-Bass.

Seeley, H. and Urquhart, C. (2008) 'Action research in developing knowledge networks', *Health Informatics Journal*, 14: 279–96.

Seibert, K.W. and Daudelin, M.W. (1999) *The Role of Reflection in Managerial Learning.* Westport, CT: Quorum.

Senge, P. (1990) *The Fifth Discipline.* New York: Doubleday.

Senge, P., Roberts, C., Ross, R., Smyth, B. and Kleiner, A. (1994) *The Fifth Discipline Fieldbook.* London: Brealey.

Shani, A.B. (Rami) and Docherty, P. (2003) *Learning by Design.* Oxford: Blackwell.

Shani, A.B. (Rami) and Pasmore, W.A. (1985) 'Organization inquiry: towards a new model of the action research process', in D.D. Warrick (ed.), *Contemporary Organization Development: Current Thinking and Applications.* Glenview, IL: Scott, Foresman. pp. 438–48.

Shani, A.B. (Rami), Mohrman, S.A., Pasmore, W., Stymne, B. and Adler, N. (eds) (2008) *Handbook of Collaborative Management Research.* Thousand Oaks, CA: Sage.

Shepard, H. (1997) 'Rules of thumb for change agents', in D. Van Eynde, J. Hoy and D.C. Van Eynde (eds), *Organization Development Classics.* San Francisco: Jossey-Bass. pp. 181–90.

Smyth, A. and Holian, R. (2008) 'Critical issues in research from within organisations', in P. Sikes and A. Potts (eds), *Researching Education from the Inside.* Abingdon: Routledge. pp. 33–48.

Stringer, E.T. (2007) *Action Research: A Handbook for Practitioners*, 3rd edn. Thousand Oaks, CA: Sage.

Susman, G. and Evered, R. (1978) 'An assessment of the scientific merits of action research', *Administrative Science Quarterly*, 23: 582–603.

Tang, Y. and Joiner, C. (2006) *Synergic Inquiry.* Thousand Oaks, CA: Sage.

Taylor, S., Rudolph, J. and Foldy, E. (2008) 'Teaching reflective practice in the action science/action inquiry tradition: key stages, concepts and practice', in P. Reason and H. Bradbury (eds), *Handbook of Action Research*, 2nd edn. London: Sage. pp. 656–68.

Tee, S., Lathlean, J., Herbert, L., Coldham, T., East, B. and Johnson, T.-J. (2007) 'User participation in mental health nurse decision-making: a co-operative enquiry', *Journal of Advanced Nursing*, 60: 135–45.

Torbert, W.R. (1976) *Creating a Community of Inquiry: Conflict, Collaboration, Transformation.* New York: Wiley.

Torbert, W.R. (1991) *The Power of Balance.* Thousand Oaks, CA: Sage.

Torbert, W.R. and Associates (2004) *Action Inquiry.* San Francisco: Berret-Koehler.

Torbert, W.R. and Taylor, S. (2008) 'Action inquiry: interweaving multiple qualities of attention for timely action', in P. Reason and H. Bradbury (eds), *Handbook of Action Research*, 2nd edn. London: Sage. pp. 239–51.

Ury, W. (1991) *Getting Past No.* London: Business Books.

Walker, B. and Haslett, T. (2002) 'Action research in management: ethical dilemmas', *Systemic Practice and Action Research*, 15 (6): 523–33.

Weisbord, M.R. (1988) 'Towards a new practice theory of OD: notes on snapshooting and moviemaking', in W.A. Pasmore and R.W. Woodman (eds), *Research in Organizational Change and Development,* Vol. 2. Greenwich, CT: JAI. pp. 59–96.

Weisbord, M.R. (2004) *Productive Workplaces*, 2nd edn. San Francisco: Jossey-Bass.

Westhues, A., Ochocka, J., Jacobson, N., Simich, L., Maiter, S., Janzen, R. and Fleras, A. (2008) 'Developing theory from complexity: reflections on a collaborative mixed method participatory action research study', *Qualitative Health Research*, 18: 701–17.

Wheelan, S.A. (1999) *Creating Effective Teams*. Thousand Oaks, CA: Sage.

White, L.P. and Wooten, K.C. (1986) *Professional Ethics and Practice in Organization Development*. New York: Praeger.

Williamson, G.R. and Prosser, S. (2002) 'Action research: politics, ethics and participation', *Journal of Advanced Nursing*, 40 (5): 587–93.

Williander, M. and Styhre, A. (2006) 'Going green from the inside: insider action research at the Volvo Car Corporation', *Systemic Practice and Action Research*, 19: 239–52.

Winter, R. (1989) *Learning from Experience: Principle and Practice in Action Research*. London: Falmer.

Winter, R. and Munn-Giddings, C. (2001) *A Handbook of Action Research in Health and Social Care*. London: Routledge.

Young, M. (1991) *An Inside Job*. Oxford: Clarendon.

Zeichner, K. (2001) 'Educational action research', in P. Reason and H. Bradbury (eds), *Handbook of Action Research*. London: Sage. pp. 273–84.

Zuber-Skerritt, O. and Fletcher, M. (2007) 'The quality of an action research thesis in the social sciences', *Quality Assurance in Education*, 15 (4): 413–36.

Zuber-Skerritt, O. and Perry, C. (2002) 'Action research within organizations and university thesis writing', *The Learning Organization*, 9 (4): 171–9.

Author Index

Subject Index